FOOD PROCESSING and BIOACTIVE COMPOUNDS

FOOD PROCESSING and BIOACTIVE COMPOUNDS

Editor

Y.S. Reddy

2016

Gene-Tech Books

New Delhi - 110 002

ISBN 81-89729-08-X

Published by : GENE-TECH BOOKS
4762-63/23, Ansari Road, Darya Ganj,
NEW DELHI - 110 002
Phone: 41562849
e-mail: genetechbooks@yahoo.co.in

Printed at : Tarun Offset Printers
Delhi

PRINTED IN INDIA

Preface

Bioprocessing is a type of advanced manufacturing that involves chemical, physical, and biological processes employed by living organisms or their cellular components. The demand for new and improved commercial products increasingly will be met through bioprocessing. It enables the translation of research discoveries into commercial products with unique and highly desirable characteristics and offers new production opportunities for a wide range of items, including foods, energy sources, commodity chemicals and pharmaceuticals.

The area of functional foods and nutraceuticals is growing worldwide and has emerged as a major trend in the food and nutrition industry. Modern food processing technologies offer a level of specificity, predictability, and productivity that otherwise would not exist in the manufacture of food products. These technologies provide for new food process designs that are cost effective, energy efficient, product specific, and environmentally benign.

This book focuses on the role of bioactive compounds in food processing. It investigates varietal differences in levels of bioactive compounds in numerous crops, and the effects of processing technologies on retention of bioactive compounds in foods. It also examines the response to many different processing operations in regard to effects on health and nutrition.

Editor

Contents

1

Bioactive Natural Compounds

Naturopathic medicine is the use of therapies that are primarily natural, including clinical nutrition, homeopathy, botanical medicine, hydrotherapy and counselling. Food products providing major calorific contribution are usually termed macronutrients, whereas essential factors such as vitamins and minerals are termed micronutrients. Products with proven biological activity that originate from biological compounds are often described as bioactives, whereas the subgroup of products that bridge the gap between food products and drugs are also termed nutriceuticals or functional foods.

Unicellular Organisms

Probiotics

Probiotics are living microorganisms that, following ingestion, form part of the colonic flora, at least temporarily, and are used with a view to improving the health and well being of the host. Various strains of probiotics, often in combination, are usedwhich differwith regard to their metabolic and immunological activities but, unlike virulent organisms that tend to amplify inflammation, probiotics may down-regulate inflammation in the intestinal mucosa. A number of uses of probiotics in gastrointestinal diseases have been proposed, including modulation of gut mucosal immunity and the prevention and treatment of intestinal infections.

Several trails of probiotic therapy for the treatment of ulcerative colitis have given encouraging results, although they appear less promising for the treatment of Crohn's disease. Owing tomany of the reported studies suffering from low patient numbers and/or using open, rather than a blinded, study protocols, further studies are required. In addition to trials for inflammatory bowel disease, studies are ongoing examining the use of probiotics to prevent or treat infectious diarrhoea. Evidence in support of their value includes the finding that Lactobacillus strain GG shortened the duration of diarrhoea in hospitalised children from 2.5 days to 1.4 days.

Botulinum Toxin

The use of the purified form of botulinum toxin from the anaerobic bacteria Clostridium botulinum now has an established place in clinical practice, due to its ability to cause temporary striated muscle paralysis. It is used extensively by neurologists for dystonic torticolis, blepharospasm, muscle spasm following stroke and by the cosmetic surgery industry to smooth facial wrinkles.

In addition, gastroenterologists have been examining its role for the treatment of patients with idiopathic anal fissures where anal muscle spasm caused by the pain exacerbates the problem, and for the treatment of achalasia when botulinum toxin is injected directly into the endoscopically located sphincter region. Its use provides symptom relief for approx. 6 months in about two-thirds of cases with achalasia. However, for both anal fissures and achalasia, its therapeutic value is limited by high recurrence rates.

Multicellular Origin

Plant Sources of Multicellular Origin

In addition to being a source of vitamins, a major area of interest for the use of biologically active chemical components of plants, i.e. phytochemicals, is for cancer prevention. For example, several phytochemicals have been shown to reduce colonic aberrant crypt foci formation caused by administration of the carcinogen azoxymethane to rats.

The list of potential anticarcinogenic phytochemicals under study is long, but includes dithiolthiones, glucosinolates and isothiocyanates (cruciferous vegetables), coumarins and limonene (citrus fruits), isoflavones inositol hexaphosphate, protease inhibitors and saponins (soybean), carotenoids (palm oil and yellow vegetables) and allium compounds (onion, garlic and leek).

Prebiotics and symbiotics

Prebiotics are a collective term for non-digestible but fermentable dietary carbohydrates that may selectively stimulate growth of certain bacterial groups resident in the colon, such as bifidobacteria, lactobacilli and eubacteria, considered to be beneficial for the human host. Digestion-resistant short-chain carbohydrates are also referred to as non-digestible oligosaccharides or low-digestible carbohydrates.

There is much interest currently in the commercial sector in adding prebiotic supplements, particularly inulin, fructo-oligosaccharides and soybean oligosaccharides, to food products to influence colonic flora, i.e. 'bifidogenic' effects. Prebiotics have a low energy value (<9 kJ/g) and increase stool volumes, but caution has been shown in their use as fermentation and osmotic side effects, such as excessive flatus, bloating, abdominal cramps and diarrhoea, may occur in a dose-related manner. Increased fermentation of short-chain fatty acids in the colon may protect against development of colorectal cancer, and the use of prebiotics to influence this activity is therefore a major area of research.

Additional areas in which prebiotic therapy may be useful include prevention of intestinal infections and modulation of the intestinal immune response in inflammatory bowel disease. An example of a prebiotic is Chicory fructooligosaccharides. Following ingestion, ChiFos reach the colon intact and acts as a prebiotic with bifi-dogenic effects.

The specificity of ChiFos for bifidobacteria is due to these bacteria secreting a β-fructosidase, allowing them to use fructo-oligosaccharides as metabolic substrates as well as inhibiting other bacteria through acidification of the media. The potential use of ChiFos as sugar substitutes and fat replacers makes them of particular interest to the food industry. Symbiotics refers to products in which a prebiotic and a probiotic are combined. The growth proliferation of the probiotic strain will be selectively promoted if it utilises the prebiotic carbohydrate.

Aloe vera has been used as a medicine since before Roman times (Figure 1). It contains several potentially bioactive compounds, including salicylates, magnesium lactate, acemannan, lupeol, campestrol, β-sitosterol, γ-linolenic acid, aloctin A and anthraquinones. It is currently available as an ingredient in a number of healthcare products where both leaf exudate and gel are available.

Figure 1. The Aloe vera plant

Aloe vera possesses multiple activities that may have beneficial effects for gastrointestinal disease, including anti-inflammatory, analgesic and pro-healing effects. For example, inflammation produced by application of croton oil to mouse ear or rat hind paw is decreased after topical application of aloe gel

In addition, a component of Aloe vera, acemannan, has been reported to accelerate healing and reduce pain in aphthous stomatitis, and prevent stress-induced gastric ulceration in rats. The wound healing properties of Aloe vera are probably not solely a consequence of its anti-inflammatory action, as in vivo and in vitro studies suggest Aloe vera or mannin derivatives can stimulate collagen synthesis and fibroblast activity. Clinical trials of Aloe vera in inflammatory bowel disease are ongoing.

Soya foods are the predominant source of dietary isoflavones and are also a rich source of trypsin inhibitor, phosphatidylinositol, saponins and sphingolipids, all of which have potential health benefits, including cancer prevention (breast, prostate and colon) and prevention of ischaemic heart disease and regulation of the host immune system. Many of these applications are, however, at a preliminary stage and further research is required to substantiate these claims. Soya protein is currently used as part of milk feed substitutes for infants with cows' milk intolerance and may also be beneficial in decreasing symptoms of the dumping syndrome in patients following vagotomy.

Curcumin (diferuoylmethane) is a yellow pigment in turmeric (Curcuma longa) and is used widely as a spice in Indian and Thai cuisine. Curcumin exhibits a number of pharmacological effects and its reported antiinflammatory activity has been attributed to suppression of prostaglandin synthesis. Preliminary reports suggest that tumeric and its extracts may be beneficial for a variety of gastrointestinal conditions, as tumeric root extract relieved pain from biliary dyskinesia in a placebocontrolled double-blinded study and curcumin capsules improved endoscopic healing of peptic ulcers as well as improving symptoms of patients with non-ulcer dyspepsia.

Curcumin decreases proliferation of the colonic cancer cell line HT29 in a concentration- and time-dependent manner. There is currently much interest in its potential as a chemopreventative agent for colorectal cancer, as curcumin possesses both cyclooxygenase-2 ('COX-2')-selective inhibitory and pro-apoptotic activity. The latter process possibly acts through the mitochondrial pathway involving caspase-8, BID cleavage, cytochrome c release and caspase-3 activation, with Bcl-2 and Bcl-xl being critical negative regulators of curcumin-induced apoptosis.

The bael (Aegle marmelos) tree grows in tropical and subtropical countries. Various parts of the bael plant are used in Ayurveda and Unani medicine for treatment of a variety of diseases, including treatment of diarrhoea, dysentery and dyspeptic symptoms. In support of this idea, methanolic extracts of unripe fruit from Aegle marmelos decreased castor oil-induced diarrhoea inmice, possibly due to the presence of tannin and mucilaginous substances.

In addition, patients suffering from diarrhoea-predominant irritable bowel syndrome showed significantly greater improvement in symptoms when given an indigenous preparation containing Aegle marmelos and Bacopa monniere compared with placebo.

Marmelosin, isolated from the bael plant, has been reported to have anti-helminthic and anti-bacterial activity. No positive effect was seen, however, in a controlled trial examining its efficacy in clinical improvement or bacteriological cure of patients with shigellosis. Further gastroenterological interest in bael comes from the finding that oral administration of luvangetin, a pyranocoumarin isolated from the seeds of Aegle marmelos, protected against multiple models of gastric ulceration in rodents. Although the exact mechanism remains unclear, this protection was probably not mediated via prostaglandin pathways.

Garlic derivatives are commonly used by the public for a variety of conditions, including as a cholesterol-lowering agent. In relation to the gastrointestinal tract, there is currently interest in its potential anti-cancer activity due to it containing the prebiotic fructo-oligosaccharide. In addition, ingestion of garlic results in the formation of diallyl disulphide from its organosulphur constituents, causing increased tissue activities of the phase II detoxification enzymes quinone reductase and glutathione transferase.

Diallyl disulphide has been shown to decrease oesophageal cancer development caused by N-nitrosomethylbenzylamine in rats. In addition to its potential anti-cancer activity, aged garlic extract has been reported to possess immuneenhancing and antioxidant properties. Further studies are required, however, to confirm these actions, and it is important to note that excessive intake of garlic derivatives may be irritant, resulting in reddening of the gastric mucosa and loss of small intestinal epithelial cells.

Extracts of the bark of pine have been used in several parts of the world as traditional medicine. Pycnogenol is a proprietary extract of the bark of the French maritime pine tree Pinus maritime. The extract is prepared by extraction of fresh pine bark with an aqueous solution of NaCl followed by ethylacetate.

The resulting product is a mixture of at least 40 different molecules, mainly flavonoids, as monomers, such as catechin, epicatechin and taxifolin, as well as condensed polymers designated as procyanidins. These molecules have antioxidant properties and may also act as modulators of metabolic enzymes, resulting in altered macrophage function, including reduced nitric oxide production. As a number of gastrointestinal diseases, such as Crohn's disease, have been associated with dysregulation of nitric oxide production, clinical trials of products containing pine bark extract could be of interest.

Animal Sources of Multicellular Origin

In the commercial sector, particularly in countries with large dairy herds such as New Zealand, milk and other dairy-derived products are undergoing extensive investigation for novel potential bioactive constituents. Some of the main products under scrutiny are described below.

Colostrum is the first milk produced after birth and is particularly rich in immunoglobulins, anti-microbial peptides (e.g. lactoferrin and lactoperoxidase) and other bioactive molecules, including growth factors. In combination with the milk that is subsequently produced, it is important for the nutrition, growth, development and immunological defence of the newborn infant. It is produced commercially as a by-product of the milk industry and is currently available in health food stores, where it is usually marketed as a general healthpromoting agent.

There is increasing evidence, however, that it may be useful for the specific treatment of both neonatal and adult gastrointestinal disease. Colostrum contains multiple specific (antibody) and non-specific (e.g. lactoferrin) antimicrobial factors in addition to several cytokines, including interleukin (IL)-1β, IL-6, IL-10, tumour necrosis factor ('TNF')- α and granulocyte-,macrophage- and granulocyte/ macrophage colony-stimulating factors. Natural bioactive substances in colostrum also include nucleosides and nucleotides.

Both the specific and non-specific constituents of colostrum may have relevance to immune modulatory and anti-microbial activity when given to neonates or adults. Several reports support the potential of colostrum for the prevention and treatment of microbial infections acting on the hosts general and specific immune function as well as the microbe itself. For example, healthy volunteers given bovine colostrum responded to attenuated oral Salmonella typhi Ty21a vaccine administration with a greater increase in circulating specific IgA compared with controls.

However, not all trials have confirmed beneficial effects, for example, adjuvant hyperimmune bovine colostrum was ineffective in decreasing the stool frequency, duration or severity of childhood shigellosis. Several studies suggest that administration of colostrum, growth-factor-enriched colostrum or purified peptides derived from colostrum may be useful to treat a variety of gastrointestinal diseases.

Administration of commercial colostral preparations have been shown to decrease non-steroidal anti-inflammatoryinduced gut injury in rats, mice and humans and to stimulate mucosal healing of patients with inflammatory bowel disease (Figure 2). Other gastrointestinal conditions for which colostrum-derived growth factors may be useful include chemotherapy-induced mucositis, necrotising enterocolitis and short bowel syndrome.

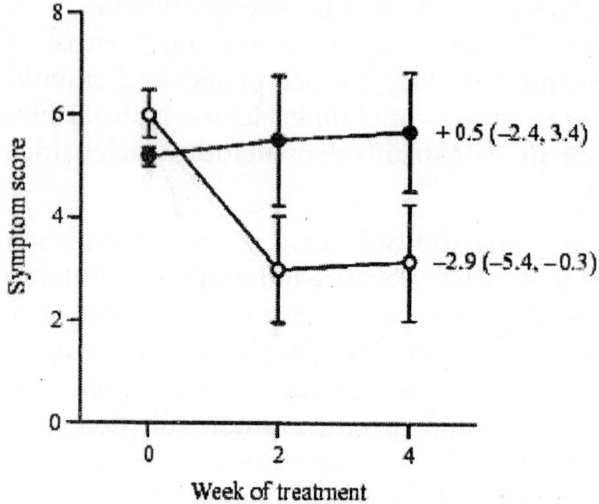

Figure 2. Colostrum therapy for the treatment of inflammatory bowel disease

The identity of the factor(s) in colostrum responsible for these biological actions are unclear. Multiple peptide growth factors are present in colostrum, including transforming growth factor ('TGF') α and β, insulinlike growth factor ('IGF') 1 and 2, epidermal growth factor (EGF) and granulocyte colony-stimulating factor ('G-CSF'). More than one peptide is likely to be important, as synergistic activity between various factors in colostrum have been demonstrated.

For example, coadministration of bovine lactoferrin with EGF resulted in synergistic activity in stimulating growth of the rat intestinal epithelial cell line IEC-18. However, two factors present in colostrum that are of particular interest for the treatment of gastrointestinal disease include transforming growth factor β and EGF. It is also important to note that there is marked variation in the relative abundance of the growth factor constituents across species (e.g. human colostrum has much higher concentrations of EGF than bovine colostrum) and when the colostrum changes to milk.

Milk products

The *caseins* are a family of phosphoproteins synthesised in the mammary gland in response to lactogenic hormones and other stimuli and secreted as large colloidal micelles. The American Dairy Science Association Committee on Nomenclature and Classification proposed bovine caseins be designated α_{s1}, αs2, β and κ caseins, with α_{s1}-casein being the major protein fraction of bovine milk. Casein contains opiod-type peptides that can influence gut motility, decreasing gastric emptying in rats, in addition to possessing immune modulatory activity using in vitro and ex vivo studies.

In addition to any specific activities present in casein, an important effect of casein ingestion may be to preserve activity and aid adsorption of other biologically active peptides by acting as a competitive substrate for pancreatic proteolytic enzymes. We have demonstrated this previously for EGF in humans, and this general mechanistic process has subsequently been shown to occur in preserving the stability and/or improving the absorption of other peptides, such as insulin-like growth factor 1, in several species.

Whey is a by-product of cheese and casein manufacture and contains approx. 20 % of the original milk proteins. These proteins include α-lactalbumin, β-lactoglobulin, lactoferrin, lactoperoxidase, immunoglobulins, glycomacropeptide and a variety of growth factors, including the EGF-receptor ligand bettacellulin. The biological activities of some of these molecules may, however, only become apparent after acidification or partial digestion. Biological activities under investigation include immune modulation and anti-inflammatory and prohealing effects; for example, α-lactalbumin decreased gastric injury caused by ethanol in rats, although it is of note that the authors report that caseinwas not protective in this model. A small human study has also suggested a beneficial effect of whey in patients with chemicalinduced corrosive injury.

Yogurt contains substantial amounts of live lactic acid bacteria and the beneficial effects of yogurt are postulated to be due to its probiotic contents, although direct effects may also occur due to its milk constituents. There is limited, mainly animal, data suggesting enhanced immune responsiveness following yoghurt administration.

Deer Antler Velvet: Each spring, the antlers of deer grow at a remarkable rate of about 1-2 cm per day for about 120 days. Deer velvet has a long history in traditional Chinese medicine for treating kidney disease, strengthening the body and healing chronic wounds. Over the last few decades, however, commercial farming of deer antler has also emerged as a product for theWestern dietary supplement market.

Deer velvet products have been shown to induce proliferation and/or morphogenic change of a variety of cell lines, including fibroblasts and the neuronal cell line PC-12. The identity of the factors responsible for these changes are unclear, although individual factors of potential interest for bone morphogenesis have been identified.

In addition, deer antlers may provide novel immunomodulatory factors, as an ethyl alcohol extraction of antler has been shown to increase murine phagocytic activity. The rapid rate of growth at the antler tip requiresmajor remodelling and angiogenic production. There is therefore currently interest in the factors controlling such responses to develop novel anti-cancer (anti-angiogenic) products.

Marine Multicellular Origins

Fish oil capsules are widely used as a general health supplement in addition to cardiovascular prophylaxis. The two major polyunsaturated fatty acids in fish oil are eicosapentaenoic and docosahexanenoic acids. Polyunsaturated fatty acids form a major constituent of animal cell membranes and eicosapentaenoic acid is also the substrate for eicosanoid production via the cyclooxygenase pathway for 3-series prostaglandins and the lipoxygenase pathway for 3-series thromboxanes.

Rats fed on diets containing high doses of fish oil have increased polyunsaturated fatty acid composition in constituents in the cell membrane of the stomach and duodenum, along with decreased prostanoid production, supporting the idea that dietary modification with fish oils may influence membrane fluidity and cellular function. Inflammatory and immune cells are also sensitive to change according to the fatty acid composition of the diet, leading to the possibility of a link between dietary polyunsaturated fatty acid intake, inflammation and immunity.

Fish oil preparations have been proposed as potential therapies for a number of gastrointestinal conditions. Several studies have reported a corticosteroid-sparing effect of gelatine-coated fish oil capsules for the treatment of ulcerative colitis, although in a direct comparison study, fish oil supplementation was less efficacious than the commonly used 5-amino-acid salicylic drug sulphasalasine.

Trials of fish oil therapy for the treatment of Crohn's disease have provided conflicting reports, with some, but not all, studies showing potential beneficial effects. Further trials appear warranted, although there is generally a requirement for more palatable formulations of n-3 fatty acids to improve patient compliance. There is also interest in the potential value of fish oils for the treatment of cancer cachexia for conditions such as pancreatic cancer.

This unusual source of bioactive compounds has been under investigation for many years, mainly in the area of development of novel anti-inflammatory drugs. The lead compound of marine origin with anti-inflammatory property has been manoalide isolated from the sponge and developed for its anti-inflammatory properties.

Manoalide has an unusual sesterterpenoid structure that antagonises phorbol-induced inflammation but not that induced by arachidonic acid, suggesting that manoalide acts prior to the cyclooxygenase step in prostaglandin synthesis, possibly by inhibiting phospholipase A2. However, it also possesses other activities, such as acting as a calcium channel blocker.

The most promising of the marine anti-inflammatory agents to date is contignasterol, a novel steroid with 14 bstereochemistry isolated from the marine sponge Petrosia contignata. This has been shown to decrease antigeninduced bronchial responsiveness in guinea-pigs, and the potential application of such anti-inflammatory drugs in gut inflammation may merit further exploration. Sponges and snails, along with severalmarine fungi and bacteria, are under active investigation for the isolation of novel anti-microbial factors.

Examples include the finding of a novel glycoside anti-fungal agent from the marine sponge Siliquariaspongia japonica and an anti-microbial peptide obtained from the marine snail Telescopium telescopium that potently inhibits the growth of Escherichia coli.

Insects Multicellular Origin

Honey has been used as a medicine since ancient times. The high sugar content of honey makes it a very hyperosmolar preparation and this has been used to advantage as a topical preparation for skin burns and other superficial infections, acting as an antibacterial agent. For example, topical honey was shown to be effective in treating postoperative skin wounds in neonates that had failed to respond to antibiotic therapy.

Honey also has the advantage of acting as a lubricant between the damaged tissue and the overlying dressing, thereby facilitating changes of wound dressing without damaging the underlying granulation tissue. Additional specific anti-bacterial activities have been reported for honey and there has been a particular focus on honey derived from the manuka plant. It is important to note, however, that some studies have failed to show a major difference in anti-microbial activity of manuka honey when compared with standard pasture honey.

Honey preparations have been proposed as potentially useful for many conditions of the gastrointestinal tract, including periodontal and other oral diseases, dyspepsia and as part of oral rehydration therapy. In vitro studies suggest that honey possesses bactericidal activity against Helicobacter pylori, although a clinical trial of manuka honey therapy to induce Helicobacter eradication failed to show a beneficial effect.

Honey may also be useful as a constituent of oral rehydration therapy, as a clinical trial in which honey was used in place of glucose for the treatment of infants and children admitted into hospital with gastroenteritis showed significantly decreased duration of diarrhoea in the honey-treated patients.

Royal jelly is used by bees to induce development of larvae into the queen phenotype and is currently sold as a general health tonic. There is extremely limited evidence supporting any major therapeutic benefit of its use and it is of note that in susceptible subjects, ingestion of royal jelly can induce asthma and anaphylaxis.

Commercial Validity for Biological Activity

Virtually all of the bioactive compounds available commercially, including those discussed above, are taken as oral preparations. Although biological activity has often been demonstrated in vitro, it is important to appreciate that this does not equate to activity once ingested. For example, many peptide growth factors are susceptible to acid/pepsin and pancreatic proteolytic digestion and are, therefore, likely to be degraded to inactive forms soon after ingestion.

In addition, many of the receptor-binding sites, such as the EGF-receptor, are not accessible from the apical (lumenal) surface and, therefore, even if they reach the small intestine intact, this does not result in receptor activation. Similarly, for non-peptide based compounds, formal pharmacokinetic and dose-response human studies have usually not been performed and yet, despite these limitations, many commercial preparations claim biological activity based solely on in vitro work.

Furthermore, the major commercial market relates to subjects taking the product as a prophylactic against illness, rather than as a therapy, but there is minimal data supporting their value for such indications. Much of the appeal of bioactive products is the perception that their natural origin equates to a 'healthier option'. However, it is important to note that toxicological injury has been caused by the use of some products, such as Chinese herbal remedies, in addition to inducing specific immunological allergic reactions.

In addition, current regulatory authority requirements for screening and post-marketing surveillance are much less stringent if a product is classified as a food supplement, rather than a drug, despite some of these compounds containing poorly identified factorswithmajor biological activity. Additional problems include potential batch variation and quality control issues and these aspects are particularly relevant if the active ingredients of the products are poorly defined. In the U.K., many of the nutriceuticals, such as colostrum, are regulated by standard food hygiene regulatory mechanisms.

In addition, the Medicines Control Agency has recognised the need for vigilance by incorporating adverse reactions reporting on unlicensed herbal remedies into their drug reaction monitoring function. Similar concerns have been raised in France, requesting that European-wide legislation should be introduced regarding the testing of bioactive products, including introducing formalised bacterial and chemical standards.

Further approaches that are currently under scrutiny are to alter the volume and nature of the components of bioactive plant and animal products. Examples include the production of genetically modified rice to provide additional vitamin A (golden rice) and using specific promoters, such as that from the β-lactoglobulin gene, to cause cattle to produce milk containing additional peptides, including human homologues of growth factors, that may be beneficial for patient treatment.

2

Bioactive Natural Products

Enzyme Inhibitors

Acetylcholinesterase Inhibitors

The principal role of acetylcholinesterase is termination of the nerve impulse transmission at the cholinergic synapses by rapid hydrolysis of the neurotransmitter acetylcholine. Bioassay-directed phytochemical investigations on a number of medicinal plants of Pakistan and Iran have led to the isolation of acetylcholinesterase inhibitors,1to 4.

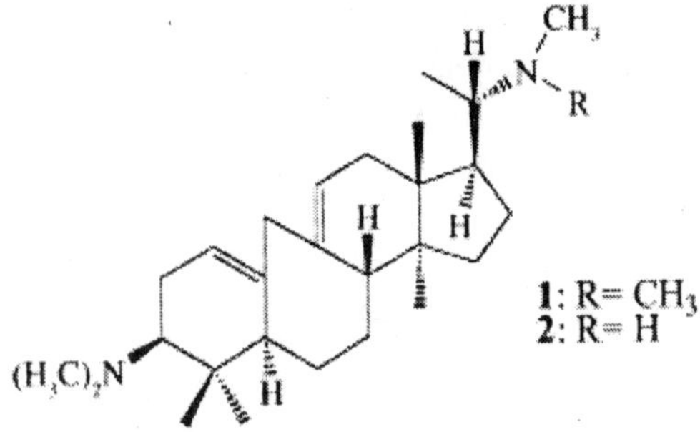

N_a,N_b-Dimethyl buxapapine (1) (IC_{50} 7.28 ± 0.06 μM)
(2) (IC_{50} 7.56 ± 0.008 μM)

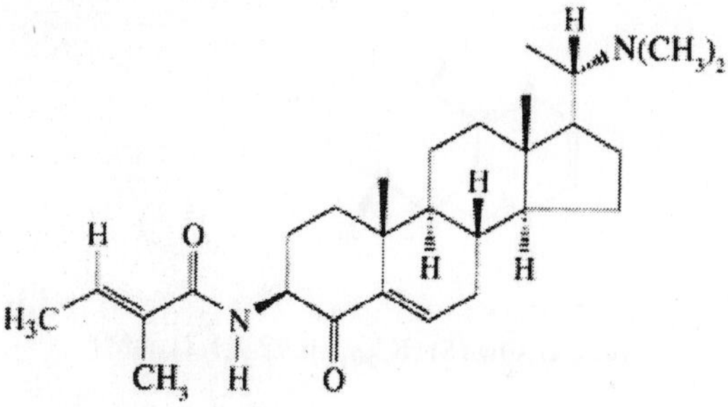

Sarsalignone (3) (IC_{50} 7.028 ± 0.007μM)

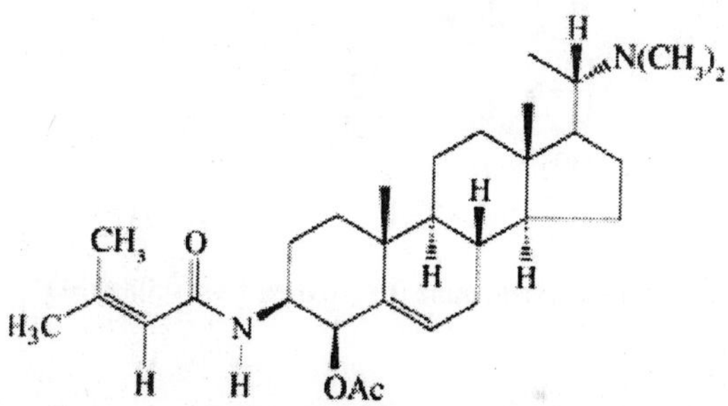

Vaganine (4) (IC_{50} 8.59 ± 0.155 μM)

Urease Inhibitors

Urease occurs in many bacteria, several species of yeast, and a number of higher plants. The enzyme catalyzes the hydrolysis of urea into CO_2, H_2O, and N_2. Our research on the development of new and safer urease inhibitors developed from natural sources has led to the isolation of a number of new urease inhibitors. The methanolic fractions of the roots of Delphinium denudatum, Sophora mollis and of stem bark of Mammea africana have yielded compounds 5-7, which showed inhibitory activity against the enzyme urease comparable to standard inhibitors.

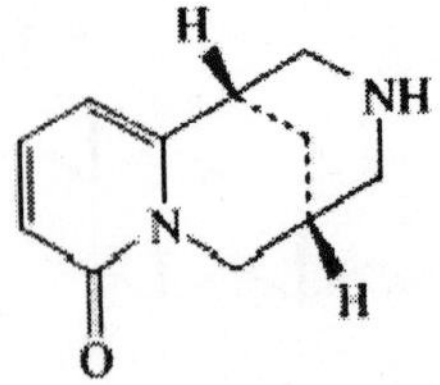

(-)-Cytisine (5) (IC_{50} = 0.92 ± 1.21 mM)

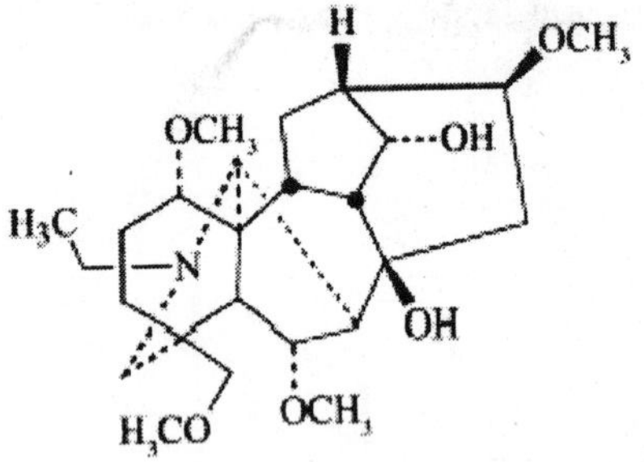

(+)-Chasmanine (6) (IC_{50} = 0.8 ± 0.03 mM)

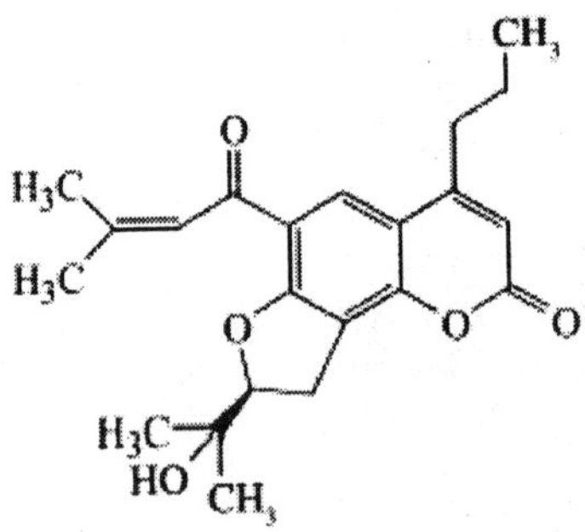

8,9-Dihydro-hydroxy-8-(1-hydroxy-1-oxobutyl-2H-furo[2,3-H]-1-benzopryan-2-one(7)(IC_{50}=60± 0.192M)

α-Glucosidase inhibitors

α-Glucosidase inhibitors delay digestion of complex carbohydrates by acting as competitive inhibitors of the intestinal enzyme α-glucosidase that hydrolyzes oligosaccharides into monosaccharides. α-Glucosidase inhibitors can hence be used to reduce the postprandial glycemic excursions and decrease postprandial hypoglycaemia.

We have identified a series of α-glucosidase inhibitors from a variety of natural sources. These include 3α-hydroxy-12-ursen-28-oic acid (8) and 3α-(4-hydroxy cinnamoyl)-12-ursen-28-oic acid (9) from the methanolic extracts of Mimusops elengi Linn.

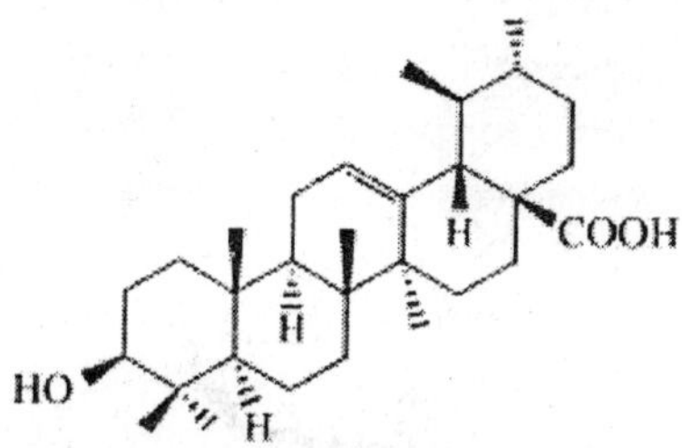

3β-Hydroxy-12-ursen-28-oic acid (8) (IC_{50} = 8.19 μM ± 0.123)

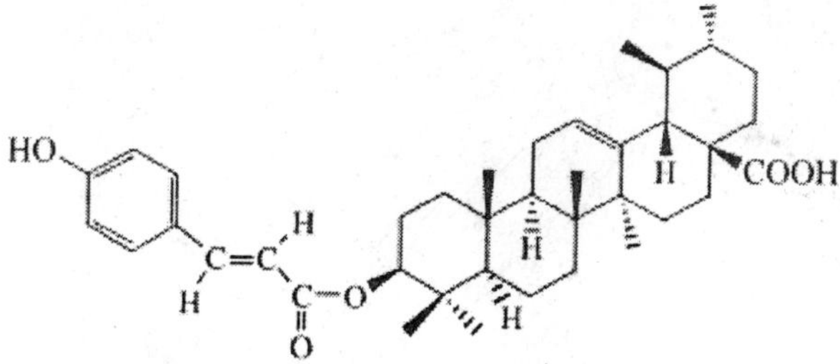

3β-(4'-Hydroxycinnamoyl)-12-ursen-28-oic acid (9) (IC_{50} = 11.65 μM ± 0.395)

Antioxidants

Free radicals such as superoxide, hydroxyl, peroxyl, and alkoxyl are produced through oxidative processes within the body by environmental pollutants, e.g., cigarette smoke, pesticides, smog, UV radiation, etc. Once formed, free radicals can damage both the structure and function of a cell membrane in a chain reaction leading to degenerative diseases and conditions such as aging processes, cataracts, acute liver toxicity, cardiovascular diseases, inflammation processes, and DNA damage that can lead to carcinogenesis.

Oxidation is also responsible for the deterioration of various organic materials ranging from the biologically important materials (e.g., lipids, foods, and oils) to the industrially important ones (e.g., rubber and lubricants). By using a battery of in vitro antioxidant assays (DPPH radical scavenging, xanthine oxidase inhibition, etc.) we have identified compounds 10-12 as potent antioxidants from natural sources.

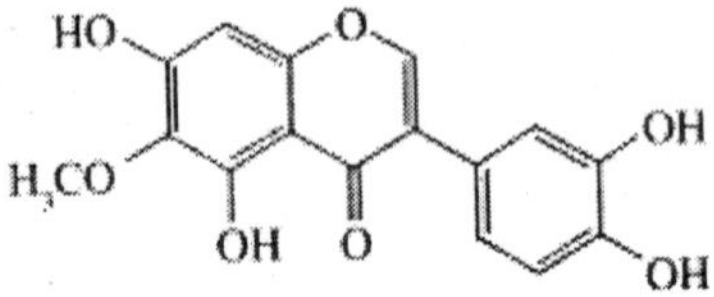

4-*H*-1-Benzopyran-4-one-3-(3',4'-dihydroxyphenyl)-5,7-dihydroxy-6-methoxy (10) (89.5% at mM)

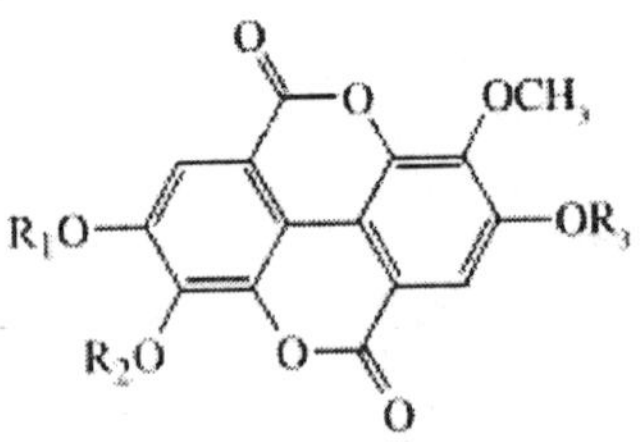

R_1,R_2= CH_2; R_3= Glu: 3,4-Methylenedioxy-3'-*O*-4-glucoside ellagic acid (11) (IC_{50} 1.096 ± 0.066 mM)

Chymotrypsin Inhibitors

Methyl[2-methoxycarbonyl] succinanilate (13), a natural aromatic amide, was isolated for the first time by us from a dark brown alga, Jolyna laminarioides Guimaraes (syn. Endrachne binghamiae).

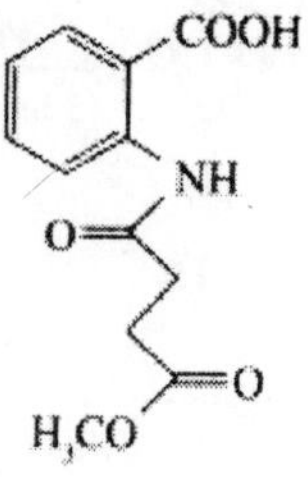

13

The chymotrypsin inhibitory activity of 13 (IC_{50} 1.14 mM) has also been studied. Chymotrypsins are serine proteases that play critical roles in several physiological processes including digestion, blood coagulation, complement activation, fibrinolysis, and reproduction.

Serine proteases are not only a physiological necessity, but can be of potential hazard if present in excessive quantities. For example, blood coagulation serine proteases are responsible for vascular clotting and cerebral and coronary disorders. The excessive activity of this enzyme can also cause skin diseases, glomerulonephritis, pancreatitis, and other disorders that can be controlled by its inhibition. Compound 13 occurs in alga in very low (i.e., 0.0002%) concentration. Therefore synthesized compound 13 for the detailed biological studies by a concise and convenient synthetic route that yielded the desired product 13 in two steps in 97% yield.

The commercially available anthranilic acid (14) was condensed with succinic anhydride (15) by refluxing in dry toluene. The desired condensed product, succinyl anthranilic acid (16), was obtained in 85% yield which was then esterified with diazomethane in etheral solution at room temperature to afford 13 in an average yield of 97% (showing in the below diagramme).

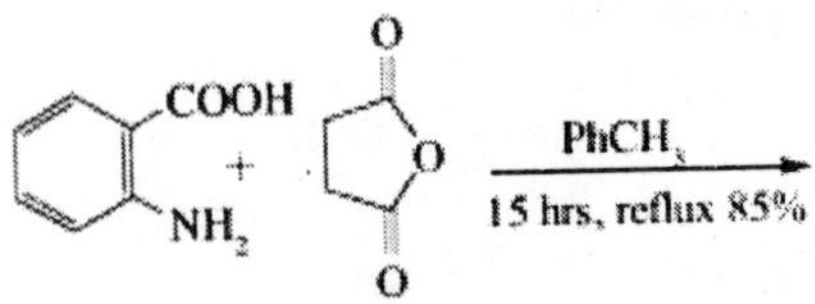

14 15

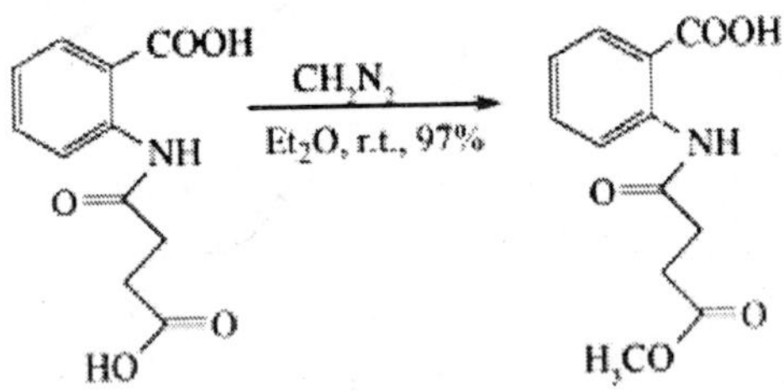

16 13

While a considerable amount of work has been done on the process of long-term potentiation , the basis of memory storage at the molecular level remains a mystery. The process of proton transfer plays a key role in many chemical and biological reactions. Intramolecular proton transfer in the transition state can play a catalytic role in some reactions and may also influence the physical and chemical properties of compounds.

A collective proton transfer can also be important in a number of biological processes such as vision and transportation in biological membrane. The mechanism of proton transfer in hydrogen bonds is related to the proton potential function which is a multidimensional surface depending on the stretching and bending modes of the hydrogen bonds and the hydrogen bond vibrations. This theory of memory depends on the orchestrated formation and breakage of hydrogen bonds across the glycoprotein molecular surfaces in the human brain.

Here present a theory of memory storage based on hydrogen bonding. It is proposed that memory is stored as a specific pattern of hydrogen bonds and other noncovalent interactions involving the hydroxyl groups present in the sugar moieties in the glycoproteins other biochemicals in the human brain. The ease with which hydrogen bonds form and break, as well as the vast, almost infinite, variation in structural possibilities which present themselves through the conformational mobility of sugar moieties in the glycoproteins present in the human brain offer an excellent template for information storage.

Hydrogen bonding can freeze sugar molecules in various conformations. The registration of an image in the human brain may correspond to the ability of the human brain to freeze an array of sugar molecules through a controlled and directed process of hydrogen bondings between sugar moieties in glycoproteins, thereby creating images and thought channels. This spectacular and fluid media of hydrogen bonds and other noncovalent interactions can hence provide the requisite powerful matrix for rapid memory storage.

The strength of these interactions, represented by the cumulative sum of individual hydrogen bonds, may thus correspond to the strength of the memory storage. The formation of covalent bonds in long-term memory processes is also conceivable. It is noticeable that hydrogen bonds play a crucial role in life processes. The conformational restriction which can be rapidly imposed through the process of hydrogen bonding in glycoproteins offers the basis of memory storage through the corresponding patterns which can result from such restrictions. The visual recognition of objects may involve the instantaneous formation of patterns in the human brain through the process of hydrogen bonding in the substrate glycoprotein structure.

3

Bioactive Substances of Plant Origin

Bioactive components of food which are of special interest include the following groups: polyphenols, phytoestrogens, phytosterols, phytates, lectins, oligosaccharides and polyunsaturated fatty acids (PUFA). These groups consist of many related compounds, each with slightly different properties. It is important to stress that the protection against cancer and cardiovascular disease is undoubtedly the result of the cumulative action of many natural substances present in the diet. Since each plant contains different bioactive components, the eating of various foods seems to be important but needs further evidence.

Bioactive Components of Food

Polyphenols

Polyphenolic compounds are mainly found in fruits and vegetables and are one of the most important sources of bioactive components of the human diet Over 8000 polyphenols have been identified and among them more than 2000 are found in nature. Plants need them for pigmentation, growth, reproduction, resistance to pathogens and for many other functions.

One of the most important groups of polyphenols is flavonoids. They can be divided into the following subgroups: flavones/flavonones, anthocyanins and catechins/flavonols. In plants, flavonoids usually form complexes with various sugars which are called glycosides. Flavones/flavonones have been isolated from almost all fruits and vegetables with their highest concentrations being found in the outer layers. Therefore flavonoid consumption can be dramatically reduced if the peel of an apple is removed.

However, there are fruits like oranges for example that have high amounts of flavonoids also present in the pulp. In most European countries, the average daily consumption of flavones/flavonones does not exceed 25 mg per day. Anthocyanins are the largest group of water-soluble pigments in plants. They are widely distributed in the human diet through crops, beans, fruits, vegetables and red wine. Researchers showed that anthocyanins can inhibit the formation of the nitrated tyrosine and scavenge peroxynitrites.

Moreover anthocyanins express a potent antioxidant activity and protective effect against hepatic ischemia-reperfusion injury in vivo. Catechins are unique flavonoids found in large quantities in green tea. In black tea the level of catechins is about 30% that of green tea. Green tea extracts are described as protective against experimentally induced cancer in animals. They act as strong inhibitors of the in vitro nitrosation of secondary amines and therefore lower tumor initiation.

High amounts of catechins present in tea are also found in red wine and chocolate, which may contribute significantly to the daily intake of polyphenols. It is estimated that the daily intake of polyphenols does not exceed 200 mg per day, which is relatively high compared to the intake of other antioxidant nutrients, such as vitamins E, C or A.

Polyphenolic compounds have beneficial health effects because of their antioxidant properties and their inhibitory role in the various stages of tumor development. There are cohort studies indicating a possible protective action against coronary heart disease and strokes. Polyphenols act through the scavenging of free radicals (reactive oxygen species, ROS) and therefore are considered to be powerful antioxidants.

Phytoestrogens

Phytoestrogens have become one of the more topical areas of interest in clinical nutrition. They mimic human estrogens and therefore are considered as natural selective estrogen receptor modulators (SERMs). There are two subclasses of polyphenols (isoflavonoids and lignans) isolated from various plants. The main consumable plant sources of phytoestrogens include isoflavonoids and lignans found mainly in soybeans and flaxseed, respectively. Other sources of phytoestrogens include sunflower and sesame seeds, various nuts, berries, garlic and carrots.

Plant lignans are also found in many cereals, grains, fruits and vegetables. Since Asians consume a lot of soybean products, their daily isoflavonoid intake is 25-100 mg. In western Europe isoflavonoid consumption is usually a few mg per day due to a much lower intake of soybean products. Phytoestrogens may prevent cancer in humans. In countries with high consumption of soybean products the risk for hormone-related prostate and breast cancer is much lower than in populations with a low consumption of soybeans.

Phytoestrogens also protect against bowel and other non-hormonal cancers, cardiovascular diseases and osteoporosis. Recently, Researchers found that soybean phytoestrogens genistein and daidzein act like estrogens in restoring nitric oxide-mediated relaxation in hypoxic rat pulmonary arteries and moreover, this effect is not mediated by the inhibition of tyrosine kinases. It is recommended to consume moderate amounts of phytoestrogens in their natural form as plant foods. If their intake is too high they could be potentially harmful to human health although this is unlikely to happen.

Phytosterols

Phytosterols are bioactive non-nutrient substances structurally similar to cholesterol. They exist in two forms:

— unsaturated, common, present in many plants and

— saturated, called stanols, which are found only in small amounts in cereals, fruits and vegetables.

Northern European daily consumption of phytosterols is in the range 100-400 mg and comes mainly from vegetable oils, bread, fruits and vegetables. In southern Europe the intake may be even higher as a result of high consumption of vegetable oils and nuts. The health effects of phytosterols are the result of their structural similarity to cholesterol; therefore plant sterols compete with cholesterol absorption from the intestinal tract. When typical amounts of sterols (240-320 mg) are consumed, only about 5% are absorbed from the small intestine.

Thus, the dietary intake of phytosterols causes an increased excretion of both dietary and biliary cholesterol in humans. In addition to reducing the absorption of cholesterol, plant stanols inhibit the absorption of other plant sterols. In humans, this inhibition of intestinal cholesterol absorption is accompanied by a compensatory increase in cholesterol synthesis, as reflected in the increase of serum cholesterol precursors, lathosterol and desmosterol. However, the net effect is still a reduction in serum cholesterol.

The beneficial effects of vegetable oils are due to the high amount of phytosterols which lower serum cholesterol. Moreover, the interaction between dietary fibre and phytosterols could explain why diets rich in fibre may reduce the risk of coronary heart disease. The use of a special margarine increases the dietary intake of phytosterols to 1-3g per day which is a therapeutic amount. In 1995, the Finns introduced plant stanol esters (PSE) in margarine, as dietary adjuncts to lower cholesterol by more than 10%.

Phytates

This group of bioactive substances is also called substances with antinutritional properties although this term is also appropriate for flavonoids. Phytates are present in seeds which are an important source of plant phosphorus stored there in the form of phytic acid (myo-inositol hexaphosphate acid, InsP6). The antinutritional effects of phytic acid are primarily related to the strong chelating associated with its six reactive phosphate groups. Its ability to complex with proteins and particularly with minerals has been a subject of investigation from chemical and nutritional viewpoints.

High contents of phytates are observed in cereal grains, legumes and nuts whereas in vegetables their concentrations are low. The hydrolysis of phytates into inositol and phosphates or phosphoric acid occurs as a result of phytase action or nonenzymatic cleavage, i.e. food processing. Enzymes capable of hydrolysing phytates are widely distributed in micro-organisms, plants and animals.

Phytases, naturally present not only in plant foods, but also in yeasts or other microorganisms implemented in food processing, act in a stepwise manner to catalyse the hydrolysis of phytic acid. To reduce or eliminate the chelating ability of phytate, dephosphorylation of hexa-

and penta-phosphate forms is essential since a high degree of phosphorylation is necessary to bind minerals.

In developing countries where the diet is almost wholly based on cereals and legumes, iron and zinc deficiency are frequently observed. Phytate-related mineral deficiencies are also reported in people from developed countries, such as pregnant women, infants and adolescents. There are several methods of decreasing the inhibitory effect of phytic acid on mineral absorption (germination, fermentation, soaking, autolysis and malting).

Unfortunately heat processing at home during cooking or during food manufacturing does not affect phytates. Therefore, the selection of starter cultures to improve phytate degradation is very important. In some cases, commercial phytase can be added to remove phytate, especially from infant foods. In recent years, dietary phytate has received increased attention due to its role in cancer prevention and/or therapy and its hypocholesterolemic effect. In turn, by binding an excess of free iron in the small intestine, phytates may prevent the formation of free radicals by the Fenton reaction in the colon and consequently decrease iron absorption for all who need less iron.

Lectins

In the past, the main scientific interest was focused on the toxicity of lectins, for example ricin. Nowadays, these bioactive compounds are recognised as natural components of the human diet. It has been shown that dietary lectins, which bind avidly and are endocytosed by cells of the brush border epithelium, are powerful growth factors for the gut, induce changes in its digestive/absorptive functions, modify the state of glycosylation of luminal receptors, alter the expression of genes coding for digestive enzymes, transport and structural proteins and interfere with both the bacterial ecology and the immune response of the gut to food antigens.

Furthermore, they stimulate pancreatic growth and have profound effects on the immune system. Plasma cells involved in a multiplicity of immune functions express high and variable levels of endogenous membrane lectins, most of which are used in cell-to-cell communication.

Oligosaccharides

Oligosaccharides represent a structurally diverse class of macromolecules of a relatively widespread occurrence in nature. They are mainly present as glucans with different types of glycosidic linkages, while others mostly bind to protein residues as oligosaccharide-protein complexes. The most promising biopharmacological activities of these biopolymers are their immunomodulation and anti-cancer effects.

Oligosaccharides and oligosaccharide-protein complexes are considered as multicytokine inducers that are able to induce the gene expression of various immunomodulatory cytokines and cytokine receptors. Numerous anti-tumor polysaccharides have been discovered from mushrooms, fungi, yeasts, algae, lichens and plants and at present are intensively studied.

Fructooligosaccharides (FOS) are short-chain polymers of fructose which are produced commercially by hydrolysis of inulin or by enzymatic synthesis from sucrose or lactose. They are not hydrolysed in the human small intestine but degraded in the colon by the resident microflora. They are mainly known for their ability to increase the endogenous growth of intestinal lactobacilli and bifidobacteria in humans and animals which is recognised as beneficial to health..

In vivo studies in rats have shown that FOS increase the proportion of butyrate which in turn stimulates water and sodium absorption and modulates intestinal motility. FOS also increase Ca, Mg and Fe absorption and enhance bone calcium stores in rats.

Polyunsaturated Fatty Acids (PUFA)

Polyunsaturated fatty acids consist of two parent compounds: linoleic acid, a fatty acid of the ω-6 family with 18 carbon atoms and two double bonds (18:2n6) and α-linolenic acid, a fatty acid of the ω-3 family with 18 carbon atoms and 3 double bonds 18:3n3). These fatty acids have different metabolic effects. Linoleic acid (LA) can be elongated to arachidonic acid (AA), a fatty acid with 20 carbon atoms and 4 double bonds (20:4n6) with two intermediary metabolites termed γ-linolenic acid (18:3n6, GLA) and dihomo-γ-linolenic acid (20:3n6, DHLA) while α-linolenic acid can be elongated to either eicosapentaenoic acid (EPA), a fatty acid with 20 carbon atoms and 5 double bonds (20:5n3) or docosahexaenoic acid DHA), a fatty acid with 22 carbon atoms and 6 double bonds (22:6n3).

The fatty acids with 20 carbon atoms, AA and EPA play an important role in prostaglandin metabolism and may influence the thrombotic process. Cohort studies indicate that the intake of α-linolenic acid is inversely related to coronary heart disease. Whether this effect is independent of other unsaturated fatty acids e.g. linoleic acid, is difficult to establish because different unsaturated fatty acids are present in the same foods e.g. soybean oil.

However, the hypothesis of a protective effect of a-linolenic acid in relation to coronary heart disease is supported by the results of the Lyon trial. In this intervention study, a Mediterranean diet enriched with a-linolenic acid was strongly protective in relation to coronary heart disease. However, more data is needed before definite statements can be made about the possible protective effect of a-linolenic acid.

A number of human intervention studies concerning antioxidant and/or anti-genotoxic effects of various polyphenols have shown no convincing results. Also extensive studies with orally administered high rates of plant extracts rich in phenolics have failed to demonstrate antioxidant effects, except for a transient improvement in the amount of trapped free radicals.

According to estimated daily intake of flavonoids, the range of 100-200 mg.day-1 is a very low level when compared with the levels used in the aforementioned studies. Recent work has shown that the bioavailability of potential antioxidants is too low to have any substantial direct effect on reactive oxygen species (Fig. 1). Moreover, the vast number of phenolics present in foodstuffs are glycosides and the free radical scavenging activity decreases with the presence of a sugar moiety, so glycosides are not antioxidants, although their corresponding aglycons are.

However, it is well recognised that many polyphenols that do not show antioxidant effects show antiinflammatory, anti- or pro-estrogenic, antimutagenic and anti-carcinogenic effects. Therefore, it is postulated that dietary compounds, even in very low concentrations, may have a far greater impact than previously anticipated, most likely by the regulation of gene expression.

This in turn can affect cellular metabolism with profound effects on detoxification mechanisms and cell proliferation, differentiation, survival and death. On the contrary, deleterious effects polyphenolic compounds have also been observed, and are associated with the ability to bind and precipitate macromolecules including protein and carbohydrates and reduce the digestibility of food. The colouring pigments in plants called flavonoids are the best known and best characterised these groups.

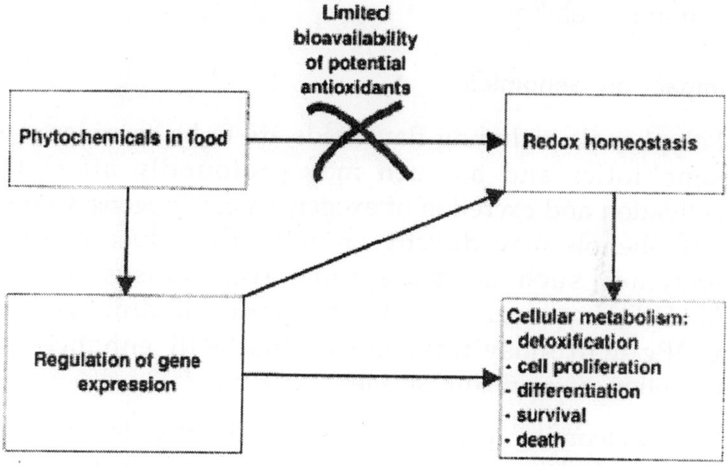

Figure 1. Relationship between bioavailibility of potential antioxidants and cell functions.

Flavonoids have been considered antinutrients because they have been shown to inhibit the activity of a wide range of enzymes including digestive enzymes such as hydrolases, but also isomerases, oxygenases, oxidoreductases, polymerases, phosphatases, protein kinases and amino acid oxidases. Failure to scavenge free radicals in vitro does not necessarily mean that some flavonoids will not trigger biological effects in vivo.

Flavonoids might interfere with various transduction signal cascades by affecting the eicosainoid synthesis via cyclooxygenase/lipoxygenase pathways, or protein tyrosine kinases. Some of them may also form complexes with metal cations, thereby interfering with the absorption of minerals such as iron or copper. The ability to bind minerals may be beneficial in some cases, since copper and iron are the initiators of hydroxyl radical formation by the

Fenton reaction. Except in extreme cases, undernourishment in western societies may actually lead to beneficial effects such as the prevention of obesity and genomic stability.

Impact on Genomics

Polyphenols including flavonoids are to be considered as xenobiotics and as such may profoundly affect the activation and excretion of exogenous carcinogens. Certain polyphenols may directly or indirectly induce phase II enzymes such as glutathione transferases (GSTs), NAD(P)H:quinone reductases, epoxide hydrolases, and UDPglucuronosyltransferases that will enhance the excretion of oxidising species.

Concomitantly, flavonoids significantly decrease the activity of antioxidant enzymes glutathione reductase (GR), catalase (CAT) and glutathione peroxidase (GPx) in the red blood cells of rats. They are also reported to influence the expression and the activity of cytochrome P450 (CYP). Antioxidant activities have shown little or no relationship to the above-mentioned antimutagenic/anticarcinogenic activities of flavonoids. How do flavonoids induce such numerous and multidirectional modifications in the intracellular biochemical apparatus? It seems likely that the effects of some of them may indirectly occur through the action on response elements in the regulatory regions of the genes.

The xenobiotic response element (XRE) is localised in the promoter regions of several genes encoding proteins such as xenobiotic metabolising enzymes (XME) of phase I (CYP1A1, CYP1A2), and phase II (NADP(P)H-menadione oxidoreductase, aldehyde dehydrogenase, UDPglucuronosyl- transferase, glutathione S-transferase) xenobiotic elimination. There are specific transcription factors, which in turn specifically bind to XRE. Once activated by the assembly with aromatic hydrocarbons or halogenated derivatives such as dioxin (TCDD), the cytosolic protein called the aryl hydrocarbon receptor (AhR) translocates to the nucleus where it heterodimerises with the aryl hydrocarbon nuclear translocator forming a transcription factor that binds to the XREs present in the 5′-promoter.

Dietary flavonoids are ligands of the AhR and affect *cyp1a1* expression, with quercetin being a very potent activator whereas kaempferol and green tea polyphenols (GTP) - catechins; especially (-)-epigallocatechin gallate (EGCG) inhibit cyp1a1 transcription induced by TCDD. Alternatively, flavonoids are reported to act through the transcriptional regulation of genes by directly affecting the antioxidant/ oxidant response element (ARE/ORE) in the promoter regions of some of the genes (gsta1, cyp1a1, cyp1a2) of the XME. It should be noted that ARE/ORE is the electrophile response element, so flavonoids may act directly on ARE/ORE as phenolic radicals or indirectly by the effects on oxidative stress.

Flavonoids have been observed to repress intrinsic antioxidant systems as a feedback mechanism exerted on antioxidant enzymes eventually pointing to the importance of intracellular prooxidantantioxidant homeostasis. There are promoter regions of several genes (including XME) that posses another response element that is activated by glucocorticoid and glucocorticoid-like structures. This, termed the glucocorticoid response element (GRE) is induced either by the glucocorticoid receptor-ligand transcription factor or by the glucocorticoid receptor-independent mechanism.

There is also the possibility that transcription factors formed by the glucocorticoid receptor-glucocorticoid interaction influence XRE the regulation of gene expression. Flavonol quercetin has been reported to selectively inhibit GREdependent gene regulation.

The activation of xenobiotic response elements by dietary agents has been known for some time, but the effects driven by the antioxidant response elements and through nuclear transcription factors such as AP-1 and nuclear factor kappa B (NF-κB) families are only now being recognised.

One important mechanism of regulation appears to be the inhibition by dietary agents of one or more of the kinase families of enzymes involved in the respective signal transduction. This may occur either by direct inhibition of the kinases itself or via the redox sensitivity of the kinase protein. At present, little is known about the molecular mechanisms of specific genes coding for proteins responsible for the observed beneficial health effects of flavonoids.

The candidates to play the key role in the regulation of cell life and death at the transcriptional level includes NF-κB and AP-1 transcription factors. Conflicting data describe the effects of catechins and teaflavins on the activation of extracellular signal-regulated kinase (ERK2) and c-Jun N-terminal kinase (JNK1) and the expression of c-jun and c-fos mRNA as well as the activity of the activator protein 1 (AP-1) (Fig. 2).

There is also contradicting evidence for the induction of AP-1 and NF-κB by the commercially used phenolic antioxidants butylated hydroxyanisol (BHA) and t-butylhydroquinone (tBHQ). They either activate NF-κB (measured by the electrophoretic mobility shift assay) with the formation of H_2O_2, or phenoxyl radicals and/or their derivatives or inhibit NF-κB DNA binding. Phenolics trigger c-Jun N-terminal kinase (JNK1) and/or extracellular signal regulated protein kinase (ERK2) in a dose-dependent fashion.

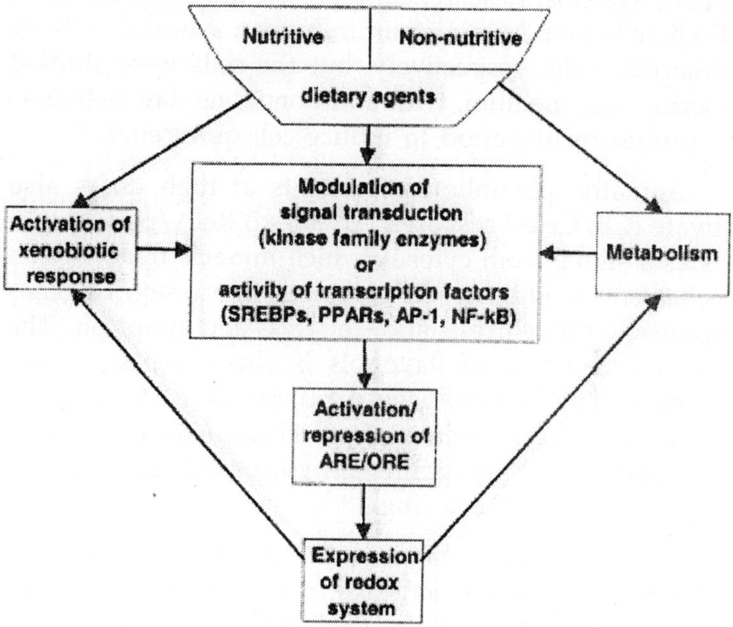

Figure 2. Metabolic pathway of nutritive and non-nutritive dietary agents.

Since JNKs are strongly and preferentially activated by stress stimuli, this signaling pathway as one of the stress responses and is functionally involved in cellular survival and/or apoptosis. It is probable that the effect of quercetin is also dose-dependent on the regulation of MAPKs and leads to the induction/repression of gene expression and cell survival or cell death.

At certain levels quercetin might be an indirect NF-κB inducer by targeting several kinases (i.e. MAP kinases), which activate NF-κB. Upstream activators include NIK, MEKK1, MEKK2, MEKK3, TAK1, protein kinase Cζ, and S6 kinase. Similarly, contradicting results were obtained from studies with quercetin and other phenolics on the activity of NF-κB, the key regulator of cellular antioxidant defence systems. Researchers reported the suppression of NF-κB activation by quercetin in human synovial cells, or glomerular cells, respectively, but the cells were studied in serum free medium, with additional one day fasting as the pretreatment period to induce cell quiescence.

Actually, phenolic antioxidants at high doses also activate ICE/Ced-3 caspases. An NF-κB RelA/p65 subunit acts as a signal from cytosol, which initiates transcription mechanisms leading to the enhanced expression of genes responsible for cell resistance to stress and apoptosis. The chemical structure of flavonols is characterised by the presence of the 2-phenylbenzen-γ-pyron ring. The ultimate difference between quercetin and kaempferol is confined to the presence of an additional hydroxyl (OH) residue in the 3′ position of the B ring.

Thus, one who compares the effect of quercetin or kaempferol on colorectal cancer cells might easily distinguish the role of the hydroxyl group present or absent in the particular flavonol. Anticarcinogenic properties of flavonols resulted in part from the inhibition of NF-κB activity. NF-κB is a ubiquitous regulator of transcription in almost every cell and it modulates the activity of genes that are characterised by the presence of the NF-κB consensus sequence in the regulatory (enhancer/promotor) regions of DNA.

Activation of NF-κB has been reported to suppress cell death, while the blockade leads to the amplification of the cytotoxic effects of TNF-α and promotes apoptosis. When stimulated, NF-κB promotes transcription, whereas the inactive form resides within the cytoplasm, blocked by the IκB subunit (with the exception of lymphocytes - B cells, where NF-κB is constitutively expressed in the nucleus).

According to the differences in the structure, at least five isoforms of NF-kB have been identified but the most abundant are subunits p50 and p65 that form homo- or heterodimers which can bind to DNA. On the contrary, TNF-a is a proinflammatory cytokine, which is known to induce cell death. TNF-α acts on the cell by the activation of the membrane receptors TNF-R1 or TNF-R2. Association with the receptors leads to conformational changes (oligomerisation into trimers) and the receptors are able to recruit a signaling complex called the DISC (death-initiated signaling complex) composed of the TRADD (TNF-R1-associated death domain) and FADD (Fasassociated death domain).

Simultaneously or alternatively, conformational changes in the receptors can recruit a signaling complex composed of TRADD and TRAF2 (TNF-associated factor 2) and/or RIP (receptor interacting protein) for survival. After the association, TRAF2 and RIP could activate kinase NIK (NF-κB inducing kinase) that stimulates NF-κB indirectly by the activation of IκBα or -β kinases (IKKα, IKKβ). IKKβ kinase i.e. phosphorylates the IκBα inhibitory subunit in positions 32 and 36 of serine residues.

After phosphorylation IκBα can be ubiquitinated and proteolytically degraded by the proteasome. Thus, NF-κB is not sequestered any more and translocates into the nucleus. NF-κB activation improves cell survival whereas inhibition enhances cytopathic and apoptotic effects of TNF-α indicating a considerable role of functional NF-κB in cell viability. The protective effect of NF-κB is in turn dependent on mRNA and proteins that regulate the activity of antiapoptotic genes. Obviously, NF-κB directly activates Bcl-2/A1 (a homologue of Bcl-2) the protein that plays an important role in the blockade of apoptosis associated with the activity of the mitochondria. Additionally, the activation of TNF-R1 by TNF-α is associated with the increased activity of PI-3K and PKB, the most powerful antiapoptotic kinases.

On the contrary, the inhibition of NF-κB was shown to occur as a consequence of proteasome inhibitors, corticosteroids, and factors that are known to block NIK and IKK, as well as under the influence of STAT-1 (signal transducer and activator of transcription 1), a tyrosine kinase that appears to be a component of the signaling complex of TNF-R1 and TRADD. Apparently, STAT-1 is recruited by TNF-R1 and enables the formation of the DISC complex; furthermore it reveals apoptotic domains of death-mediating proteins with the concomitant inhibition of the assembly of the survival complex, which releases NF-κB.

Substantial interest to study the physiological role of the STAT-1 resulted from its dual role as the non-receptor tyrosine kinase and transcription factor (the STAT acronym stands for that meaning). One should bear in mind, however, that flavonoids are known as the most powerful inhibitors of tyrosine kinases. Whether flavonols are potent inhibitors/activators of STAT-1 and effect signal transduction form TNF-α to NF-κB with simultaneous activation of genes supporting cell viability, is a matter of debate and needs experimental verification which is currently in progress in our laboratory.

The developing resistance of tumor cells to chemotherapy is a challenge to contemporary medicine. A number of drug resistance mechanisms are not known as well as the origin of this phenomenon. Apparently, apoptosis is a hallmark of an efficient cytotoxic effect of chemotherapy or radiation therapy. Apoptosis is also widely accepted as a mechanism leading to cell elimination induced by TNF-α. Therefore, TNF-α either/ or chemotherapy as well as radiation therapy might be considered as important activators of NF-κB. It has been demonstrated that inhibition of NF-κB supports the therapies based on the action of TNF-α.

A marked role of NF-κB has been observed in developing resistance to chemotherapy in the following cases: Hodgkin lymphoma, juvenile myelomonocytic leukemia, prostate cancer, virus-mediated leukemic T cells and tumor cells transformed with Ras oncogene. Tumor cells are also characterised by a higher nuclear representation of NF-κB, moreover the genes regulated by NF-κB are often constitutively upregulated in neoplasmas. NF-κB is thus linked to tumor growth, because it inhibits apoptosis. Several experimental data support the evidence of a profound role played by NF-κB in TNF-α-mediated apoptosis.

Overexpression of IκB renders tumor cells significantly less susceptible to TNF-α-induced cell death. The reaction was observed in tumor cell lines such as Jurkat T cells, human urine bladder line T24 and breast cancer MCF7. On the contrary, flavonoids are cytopathic to tumor cells. The puzzling issue of the developing resistance of tumor cells to cytopathic actions of flavonols in the presence of TNF-α-stimulated NF-κB activity remains unexplained and ambiguous.

Another possible mechanism of cell toxicity of flavonoids seems to be quite similar to that reported by some researchers for other phenolic compounds such as gallic acid and its esters, which inhibit protein tyrosine kinases (PTKs). Similar findings were obtained by another researchers on the basis of studies with quercetin and resveratrol in cultured rat stellate and Kupfer cells. In their studies, the action of quercetin aglycon was associated with suppressed inositol phosphate metabolism, tyrosine phosphorylation, reduced level of cell cycle protein cyclin D1 and mitogen-activated (MAP) kinase activation in PDGF/BB stimulated stellate cells.

In certain cases quercetin aglycon could promote tumorigenesis and tumor growth possibly by oxidative

DNA damage in the presence of Cu^{2+}. The anticarcinogenic activity of quercetin aglycon is dose dependent and is influenced by the presence of cytotoxic agents as well as serum survival factors (cytokines, oxygen free radicals - OFR). Phenolics stay in the first line of antioxidant defence, donating electrons to OFR with a resultant formation of phenoxyl radicals. Prooxidant phenoxyl radicals co-oxidise NADH and GSH, which in turn are no longer able to inhibit NF-kB activation. It appears that apoptosis may be initiated by phenolics.

According to recent advances in the interpretation of events that occur during programmed cell death, the generation and spreading of ROS within the cell are consequences of the increased permeability of the mitochondrial membrane. Lepley and Pelling during a novel cell culture study with apigenin (a quercetin derivative), obtained evidence that flavonoid antioxidants may enhance apoptosis in certain tumor cell lines. Hydrogen peroxide (H_2O_2), similarly to quercetin has been reported to stimulate the activity of the mitogen-activated protein kinases (MAPKs). ERK and JNK, and the expression of the proto-oncogenes c-fos and c-jun.

Cell elimination or survival is then a matter of balance between the protein products of genes controlling cell life and death. The key roles whether an individual cell dies or remains alive may be the modulation of antioxidant defences. Quercetin has been suggested to be beneficial for health, however, studies have shown that many antioxidants can also exhibit, a prooxidant behavior. Plant polyphenols including quercetin aglycon may interfere with the cellular redox state by the inhibition of glutathione reductase and the efflux of glutathione S-conjugates.

The final effect of plant phenolics on the viability of cells is therefore variable. At low concentrations quercetin and derivatives seem to exert a stimulatory action on cell viability and survival, whereas at high doses (100 mM and higher) they are apoptogenic and cytotoxic. From a dietary point of view, it remains to be established what doses could be considered as beneficial for health.

Role of PUFA as Ligands

The development of obesity and associated insulin resistance involves a multitude of gene products, including proteins involved in lipid synthesis and oxidation, thermogenesis and cell differentiation. The dietary ω-6 and ω-3 polyunsaturated fatty acids (PUFA) suppress lipogenesis in the liver while they simultaneously induce the transcription of genes encoding proteins of lipid oxidation and thermogenesis. Furthermore, the lipoprotein metabolic pathway is altered by peroxisome proliferatoractivated receptors (PPARs).

The PPARs are a member of the steroid hormone receptor superfamily. Three types of PPARs have been described: PPARα, PPARβ/δ (Nuc1), and PPARγ. PPARα and PPARβ are ubiquitously expressed in body tissues that predominantly catabolise fatty acids (i.e. heart, liver, muscle, brown adipose tissue) whereas PPARγ is selectively expressed in adipose tissues (recently found in other tissues including skeletal muscle) and seems to be associated with the differentiation of adipocytes. PPARα is activated by PUFA such as eicosapentaenoic acid (ω-3; EPA) or linoleic acid (ω-6, LA) and heterodimerises with the 9-cis-retinoic acid receptor (RXRα).

After ligand binding (EPA), it functions as a transcription factor in the regulation of adipogenesis and insulin-mediated glucose transport. There is a positive correlation between the expression of Cu, Zndependent superoxide dismutase (SOD-1) and PPARα. Moreover, the same group observed that an increased liver PPARa mRNA level confers a reduction of the plasma TBARS levels indicating the causative role of reactive oxygen species in the pathology of insulin resistance. PUFA are not only strong ligands to PPARα, but also strong activators of PPARγ and PPARβ. Ligand binding enhanced the interaction of PPAR with its DNA-binding domains called the PPAR response element (PRE).

Functional PRE that reside in the 5′-flanking region have been found to exist in several genes including those involved in the oxidation of fatty acids and thermogenesis (uncoupling proteins UCP-1 and UCP-3), thus additionally supporting the evidence for its anti-diabetic and antiobesity function. Repartitioning of metabolic fuels away from storage and towards oxidation reflects the fact that PUFA co-ordinately suppress the transcription of lipogenic genes, while simultaneously inducing the transcription of genes encoding proteins of lipid oxidation (β-oxidation of fatty acids) and thermogenesis. This effect of PUFA is in turn mediated by the transcriptional and translational suppression of another group of transcription factors termed sterol regulatory elementbinding proteins (SREBPs).

Thus, PUFA play a beneficial role in health by a hypolipidemic action by lowering plasma cholesterol and preventing atherosclerosis, hypertension, cardiovascular diseases, obesity and insulin resistance. This is further corroborated in studies performed by Mohan et al., who observed that oral supplementation with oils rich in ω-3 and ω-6 PUFA could protect animals against alloxaninduced diabetes mellitus. It is thought that PUFA exert the aforementioned effect by enhancing the antioxidant status and suppressing the production of cytokines (TNF-α in particular). Apparently, PUFA play a role as intrinsic ligands in activating the PPARγ - transcription factor, which up to date has been known to be activated merely by thiazolidinediones.

4

Bioactive Metabolites

Cowpea is a key staple food in many developing countries, and forms an integral part of the diet of about 120 million people around the world. All the parts of the plant that are used for food are nutritious, providing protein, vitamins and minerals. Cowpea is an herbaceous short term, annual leguminous plant which is grown in many tropical and subtropical countries. Cowpea grains contain about 25% protein, and the ability of cowpea plants to tolerate drought and poor soils makes it an important crop in the Savanna regions where these constraints restrict other crops.

However, cowpea production and storage has been hampered over the years by the menace of various pests and diseases, which attack the crop at various stages of its production. Major pests attacking cowpea plant includes flower thrips, pod borer and pod sucking bugs. Storage weevil (Callosobruchus maculatus) also damage stored cowpeas. Diseases of cowpea include stem and root rots and leaf spots caused by fungal, while mosaic disease and mottle symptoms are caused by viruses.

Consequently, scientist has sought for ways for combating the menace of cowpea pests and diseases. Various methods have been device to minimise if not eliminate the damaging effect of these diseases and pests. One of such method adopted is the development and breeding of cowpea crops that are resistant to pests and diseases. An example of which is the "Bulk Population Breeding" method in which different species of cowpea with different traits are cross breed to improve yield.

Another method of recent development is the genetic manipulation in which the required gene from wild relatives of plants related to cowpea is transferred to the plant through cross breeding or in some cases the foreign gene is inserted into cowpea plant using the soil bacterium, Agrobacterium tumefaciens as a vector. However, over the past few years through the establishment of a regional pest management, International Institute of Tropical Agriculture (IITA) has been able to launch a number of efficient pest control technologies for cowpea plant, mainly based on resistant, higher-yielding varieties. IITA has thus been able to developed high-yielding varieties for both sole and intercropping, with resistance to major disease, insect pests, nematodes, and parasitic weeds.

This research work thus aims at establishing whether or not disease and pest resistant cowpea plant produces toxic compounds which can be detected by the Brine Shrimp lethality Test (BST), so that on the long run compounds that can be used to protect crops from pests and disease can be isolated from cowpea.

Eight different cowpea cultivars, which were supplied, by Dr. Bir B. Singh and Dr. Stephen K. Asante of the International Institute of Tropical Agriculture (IITA), Kano Station, were used in this investigation. The brine shrimp (Artemia salina) eggs used is of the brand "Argentenmia" packaged by Argent Chemical Laboratories, USA, while synthetic sea salt of the brand "Instant Ocean" manufactured by "Aquarium Systems,

Sarrebourg, France" was used to prepare the sea salt solution. 50 g of grounded cowpea seeds was extracted with 200 ml of ethanol for four days, after which the mixture was filtered and the residue was re-extracted with a fresh 200 ml of ethanol for another four days and filtered.

The combined filtrate was concentrated on "Buchi rotary evaporator" at 40°C. The residue was collected, and weighed. About 40 mg of the extract was kept for testing and the remaining was partitioned between 100 ml of $CHCl_3/H_2O$ (1:1) mixture, the two distinct layers formed overnight were separated.

The $CHCl_3$ soluble fraction was evaporated to dryness and weighed, and the aqueous fraction was extracted with 50ml ethyl acetate. The ethyl acetate layer was dried with anhydrous Na_2SO_4, filtered and evaporated to dryness. The aqueous layer was discarded. Brine shrimp lethality assay was carried out for each of the fraction of all the 8 cowpea cultivars.

A compound is considered to be cytotoxic if it inhibits vital metabolic processes or it causes disorder in living organisms resulting in perversion of behavior or death. Several plant extracts are known to contain cytotoxic compounds. For example, Azadirachta indica extracts affects about 195 species of insects at concentration ranging from 0.1 to 1000 ppm. More importantly, insects that have become resistant to synthetic pesticides can be controlled with some plant extracts.

Results presented in Tables 1 - 3 shows the cytotoxic properties of the ethanolic, chloroform and ethyl acetate extracts of improved cowpea seeds at 10, 100 and 1000 μg/ml concentration. Generally the cytotoxic activity increases with increase in concentration in most of the active extracts. The extracts of IT93K - 596 - 9 - 12 are the most active.

Table 1. Brine shrimp lethality assay results for the ethanol extracts.

Sample Code	Dosage (µg/ml)	Number of Mortality	Percentage Mortality
IT93K – 596 – 9 – 12	10	4	13.33
	100	8	26.67
	1000	28	93.33
IT90K – 277 – 2	10	0	0.00
	100	3	10.00
	1000	22	73.33
IT93K – 452 – 1	10	1	3.33
	100	2	6.67
	1000	19	63.33
IT845 – 2246 – 4	10	0	0.00
	100	3	3.33
	1000	7	23.33
IT89KD – 288	10	0	0.00
	100	0	0.00
	1000	3	3.33
IT86D – 719	10	1	3.33
	10	0	0.00
	1000	0	0.00
IT93K – 389	10	1	3.33
	100	1	3.33
	1000	9	30.00
IT90K – 76	10	0	0.00
	100	2	6.67
	1000	8	26.67

For most of the cowpea cultivars, the ethyl acetate soluble fractions are most active. The ethanolic extract of IT93K - 452 - 1 recorded the highest mortality rate.

Table 2. Brine shrimp lethality assay results for the chloroform extracts.

Sample Code	Dosage (μg/ml)	Number of Mortality	Percentage Mortality
IT93K – 596 – 9 –12	10	2	6.67
	100	16	53.33
	1000	29	96.67
IT90K – 277 – 2	10	1	3.33
	100	1	3.33
	1000	9	30.00
IT93K – 452 – 1	10	0	0.00
	100	0	0.00
	1000	2	6.67
IT845 – 2246 – 4	10	1	3.33
	100	1	3.33
	1000	2	6.67
IT89KD – 288	10	1	3.33
	100	0	0.00
	1000	0	0.00
IT86D – 719	10	0	0.00
	100	1	3.33
	1000	6	20.00
IT93K – 389	10	0	0.00
	100	1	3.33
	1000	0	0.00
IT90K – 76	10	0	0.00
	100	2	6.67

	1000	8	26.67

Table 3. Brine shrimp lethality assay results for the ethyl acetate extracts.

Sample Code	Dosage (μg/ml)	Number of Mortality	Percentage Mortality
IT93K – 596 – 9 – 12	10	0	0.00
	100	20	66.67
	1000	30	100.00
IT90K – 277 – 2	10	0	0.00
	100	1	3.33
	1000	22	73.33
IT93K – 452 – 1	10	0	0.00
	100	0	0.00
	1000	10	33.33
IT845 – 2246 – 4	10	0	0.00
	100	0	0.00
	1000	24	80.00
IT89KD – 288	10	0	0.00
	100	0	0.00
	1000	29	96.67
IT86D – 719	10	1	3.33
	100	1	3.33
	1000	10	33.33
IT93K – 389	10	0	0.00
	100	0	0.00
	1000	3	10.00
IT90K – 76	10	0	0.00
	100	0	0.00
	1000	1	3.33

Jones and coworkers established that azadirachtin an extract of neem tree A. indica block the development of the motile male gamete in vitro. similarly, some other plants including Anona squamosa (atis or sugar apple), Eucalyptus globus, Lansium domesticum (lansones), and Cundiacum variegatum have been found to posses insecticidal properties from their activities against Aedes aedes aegypti (Linnaens) and Culex quinque fasciators Mazon et al. These plant extracts which posses insecticidal and pesticidal properties can be developed into insecticide and pesticides that will be environmentally safe alternatives to synthetic pesticides.

The results of the Brine Shrimp lethality Bioassay experiments indicates the presence of active cytotoxic compounds in the extracts of the seeds of the manipulated cowpea cultivars IT93K - 596 - 9 - 12, IT90K - 277 - 2 and IT93K - 452 - 1. These cultivars need to be investigated further so as to isolate and identify the compounds responsible for the activity.

5

Food Biotechnology

The first generation of transgenic crops primarily benefited farmers. Although there are consumer benefits in growing these crops, the benefits are largely invisible to consumers. For example, studies have shown that because insect-resistant corn (Bt corn) sustains relatively little insect damage, fungi and molds cannot infect those plants as easily as non-insect-resistant crops.

Therefore, the level of toxins, such as aflatoxin, produced by these pathogens, some of which are fatal to livestock, is much lower in Bt corn than non-Bt corn. The benefits of the next wave of biotechnology crops will be more obvious to consumers. Some of those benefits will involve improvements in food quality and safety, while others will provide consumers with foods designed specifically to be healthier and more nutritious.

A variety of healthier cooking oils derived from biotechnology are already on the market. Using biotechnology, plant scientists have decreased the total amount of saturated fatty acids in certain vegetable oils. They have also increased the conversion of linoleic acid to the fatty acid found mainly in fish that is associated with lowering cholesterol levels. Another nutritional concern related to edible oils is the negative health effects produced when vegetable oils are hydrogenated to increase their heat stability for cooking or to solidify oils used in making margarine. The hydrogenation process results in the formation of trans-fatty acids.

Biotechnology companies have given soybean oil these same properties, not through hydrogenation, but by using biotechnology to increase the amount of the naturally occurring fatty acid, stearic acid. Animal scientists are also using biotechnology to create healthier meat products, such as beef with lower fat content and pigs with a higher meat-to-fat ratio.

Other health and nutritional benefits of crops improved through biotechnology include increased nutritional value of crops, especially those that are food staples in developing countries. Scientists at Nehru University in New Delhi used a gene found in the South American plant amaranth to increase the protein content of potatoes by 30 percent. These transgenic potatoes also contain large amounts of essential amino acids not found in unmodified potatoes. Other examples include golden rice and canola oil, both of which are high in vitamin A. The golden rice developers further improved rice with two other genes that increase the amount and digestibility of iron.

Biotechnology also promises to improve the health benefits of functional foods. Functional foods are foods containing significant levels of biologically active components that impart health benefits beyond our basic needs for sufficient calories, essential amino acids, vitamins and minerals. Familiar examples of functional foods include compounds in garlic and onions that lower cholesterol and improve the immune response; antioxidants found in green tea; and the glucosinolates in broccoli and cabbage that stimulate anticancer enzyrnes.

We are using biotechnology to increase the production of these compounds in functional foods. For example, researchers at Purdue University and the U.S. Department of Agriculture, created a tomato variety that contains three times as much of the antioxidant lycopene as the unmodified variety. Lycopene consumption is associated with a lower risk of prostate and breast cancer and decreased blood levels of "bad cholesterol."

Quality of the Product

We are also using biotechnology to change the characteristics of the raw material inputs so that they are more attractive to consumers and more amenable to processing. Biotechnology researchers are increasing the shelf life of fresh fruits and vegetables; improving the crispness of carrots, peppers and celery; creating seedless varieties of grapes and melons; extending the seasonal geographic availability of tomatoes, strawberries and raspberries; improving the flavor of tomatoes, lettuce, peppers, peas and potatoes; and creating caffeine-free coffee and tea.

Much of the work on improving how well crops endure food processing involves changing the ratio of water to starch. Potatoes with higher starch content are healthier because they absorb less oil when they are fried, for example. Another important benefit is that starchier potatoes require less energy to process and therefore cost less to handle. Many tomato processors now use tomatoes derived from a biotechnology technique, somaclonal variant selection. The new tomatoes, used in soup, ketchup and tomato paste, contain 30 percent less water and are processed with greater efficiency.

Another food processing sector that will benefit economically from better quality raw materials is the dairy products industry. Scientists in New Zealand have now used biotechnology to increase the amount of the protein casein, which is essential to cheese making, in milk by 13 percent. Biotechnology also allows the economically viable production of valuable, naturally occurring compounds that cannot be manufactured by other means. For example, commercial-scale production of the natural and highly marketable sweetener known as fructans has long eluded food-processing engineers.

Fructans, which are short chains of the sugar molecule fructose, taste like sugarbut have no calories. Scientists found a gene that converts 90 percent of the sugar found in beets to fructans. Because 40 percent of the transgenic beet dry weight is fructans, this crop can serve as a manufacturing facility for fructans.

Safety of the Raw Materials

The most significant food-safety issue food producers face is microbial contamination, which can occur at any point from farm to table. Any biotechnology product that decreases microbes found on animal products and crop plants will significantly improve the safety of raw materials entering the food supply. Improved food safety through decreased microbial contamination begins on the farm. Transgenic disease-resistant and insect-resistant crops have less microbial contamination.

New biotechnology diagnostics detect microbial diseases earlier and more accurately, so farmers can identify and remove diseased plants and animals before others become contaminated. Biotechnology is improving the safety of raw materials by helping food scientists discover the exact identity of the allergenic protein in foods such as peanuts, soybeans and milk, so they can then remove them. Although 95 percent of food allergies can be traced to a group of eight foods, in most cases we do not know which of the thousands of proteins in a food triggered the reaction. With biotechnology techniques, we are making great progress in identifying these allergens.

More importantly, scientists have succeeded in using biotechnology to block or remove allergenicity genes in peanuts, soybeans and shrimp. Finally, biotechnology is helping us improve the safety of raw agricultural products by decreasing the amount of natural plant toxins found in foods such as potato and cassava.

Food Processing

Microorganisms have been essential to the food-processing industry for decades. They play a role in the production of the fermented foods. They also serve as a rich source of food additives, enzymes and other substances used in food processing. Because of the importance of fermented foods to so many cultures, scientists are conducting a lot of work to improve the microorganisms that carry out food fermentations.

The bacterium responsible for many of our fermented dairy products, such as cheese and yogurt, is susceptible to infection by a virus that causes substantial economic losses to the food industry. Through recombinant technology, researchers have made some strains of this bacterium and other important fermentors resistant to viral infection. Some bacteria used in food fermentation produce compounds that kill other, contaminating bacteria that cause food poisoning and food spoilage. Using biotechnology we are equipping many of our microbial fermentors with this self-defense mechanism to decrease microbial contamination of fermented foods.

Food Additives

Food additives are substances used to increase nutritional value, retard spoilage, change consistency and enhance flavor. The compounds food processors use as food additives are substances nature has provided and are usually of plant or microbial origin, such as xanthan gum and guar gum, which are produced by microbes. Many of the amino acid supplements, flavors, flavor enhancers and vitamins added to breakfast cereals are produced by microbial fermentation.

Through biotechnology, food processors will be able to produce many compounds that could serve as food additives but that now are in scant supply or that are found in microorganisms or plants difficult to maintain in fermentation systems. Food processors use plant starch as a thickener and fat substitute in low-fat products. Currently, the starch is extracted from plants and modified using chemicals or energy-consuming mechanical processes.

Scientists are using biotechnology to change the starch in crop plants so that it no longer requires special handling before it can be used. Enzymes produced by microbial fermentation play essential roles as processing aids in the food industry. The first commercial food product produced by biotechnology was an enzyme used in cheese making. Prior to biotech techniques, this enzyme had to be extracted from the stomach of calves, lambs and baby goats, but it is now produced by microorganisms that were given the gene for this enzyme.

The production of high-fructose corn syrup from cornstarch requires three enzymes, and those same enzymes are important in making baked goods and beer. Other enzymes are essential to the production of fruit juices, candies with soft centers, and cheeses. The food industry uses more than 55 different enzyme products in food processing. This number will increase as we discover how to capitalise on the extraordinary diversity of the microbial world and obtain new enzymes that will prove important in food processing.

Safety of the Food Supplay

In addition to the many ways biotechnology is helping us enhance the safety of the food supply, biotechnology is providing us with many tools to detect microorganisms and the toxins they produce. Monoclonal antibody tests, biosensors, polymerase chain reaction (PCR) methods and DNA probes are being developed that will be used to determine the presence of harmful bacteria that cause food poisoning and food spoilage, such as Listeria and Clostridium botulinum.

We can now distinguish E. coli 0157:H7, the strain of E. coli responsible for several deaths in recent years, from the many other harmless E. coli strains. These tests are portable, quicker and more sensitive to low levels of microbial contamination than previous tests because of the increased specificity of molecular technique. Biotechnology-based diagnostics have also been developed that allow us to detect toxins, such as aflatoxin, produced by fungi and molds that grow on crops, and to determine whether food products have inadvertently been contaminated with peanuts, a potent allergen.

The contributions that biotechnology has made to health care and agriculture have received much attention from the press and the public, but now society is beginning to see the benefits of biotechnology's "third wave"-industrial and environmental biotech. This third wave of biotechnology is already successfully competing with traditional manufacturing processes and has shown promise for achieving industrial sustainability. To industry, sustainable development means continuous innovation, improvement and use of "clean" technologies to make fundamental changes in pollution levels and resource consumption. An industrially sustainable process should, in principle, be characterised by

— reduction or elimination of waste.

— low consumption of energy and nonrenewable raw materials (and high use of carbohydrate feedstocks, such as sugars and starch).

Living systems manage their chemistry more efficiently than man-made chemical plants, and the wastes that are generated are recyclable or biodegradable. Biocatalysts, and particularly enzyme-based processes, operate at lower temperatures and produce less toxic waste, fewer byproducts and less emissions than conventional chemical processes. They may also use less purified raw materials (selectivity).

Use of biotechnology can also reduce energy required for industrial processes. Finally, just as biotechnology is providing us with new tools for diagnosing health problems and detecting harmful contaminants in food, it is yielding new methods of monitoring environmental conditions and detecting pollutants. Biotechnology in industry employs the techniques of modern molecular biology to reduce the environmental impact of manufacturing. Industrial biotechnology also works to make manufacturing processes more efficient for industries such as textiles, paper and pulp, and specialty chemicals.

Some observers predict biotechnology will transform the industrial manufacturing sector in much the same way that it has changed the pharmaceutical, agricultural and food sectors. Industrial biotechnology will be a key to achieving industrial and environmental sustainability.

According to the Organisation for Economic Cooperation and Development (OECD), industrial sustainability is the continuous innovation, improvement and use of clean technology to reduce pollution levels and consumption of resources. Modern biotechnology provides avenues for achieving these goals. In recent years, policymakers, corporate executives, private citizens and environmentalists have become more concerned about sustainable development. In response to that concern, many leading industrial companies are doing more than meeting their legal minimums. Many are developing policies and implementation plans for sustainability that include guidelines for environmental health and safety as well as product stewardship.

The key words to achieving sustainability are "clean" and "efficient." Any change in production processes, practices or products that makes production cleaner and more efficient per unit of production or consumption is a move toward sustainability. In practical terms, industrial sustainability means employing technologies and know-how to lessen material and energy inputs, maximise renewable resources and biodegradable substances as inputs, minimise the generation of pollutants or harmful waste during product manufacture and use, and produce recyclable or biodegradable products.

Industrial Manufacturing Processes

Manufacturing processes have long relied on petroleum, a nonrenewable resource that generates pollution and solid waste, as a source of material and energy. Biotechnology provides ways to produce cleaner products and processes by reducing the use of petroleum inputs. Industrial biotechnology instead uses natural sugars as feedstocks.

Through biotechnology, the use of renewable, biomassbased feedstocks will increase. Bio-feedstocks offer two environmental advantages over petroleum-based production: Production will be cleaner, in most cases, and less waste will be generated. When the biomass source is agricultural refuse, our gains double: We will enjoy all the advantages of bio-feedstocks while reducing wastes generated from another human endeavor agriculture. A final advantage of using plant biomass as feedstock is that as our crop of feedstock grows, it consumes CO_2-one of the greenhouse gases.

Biotechnology will also have an impact on two sources of energy: fossil fuels and new biomass-based fuels. Innovations wrought by biotechnology can help remove the sulfur from fossil fuels, significantly decreasing their polluting power. Using biomass for energy has the same environmental advantages as using biomass feedstocks, so government labs have devoted significant resources to research on recombinant technology and bioprocess engineering to improve the economic feasibility of biomass-derived energy.

In addition to working toward sustainability by using biomass-based material and energy inputs, biotechnology offers us many options for minimising the environmental impact of manufacturing processes by decreasing energy use and replacing harsh chemicals with biodegradable molecules produced by living things. Unlike many chemical reactions that require very high temperatures and pressures, reactions using biological molecules work best under conditions that are compatible with life-that is, temperatures under 100° F, atmospheric pressure and water-based solutions. Therefore, manufacturing processes that use biological molecules can lower the amount of energy needed to drive reactions.

Manufacturing processes that use biodegradable molecules as biocatalysts, solvents or surfactants are also less polluting. Microbial fermentation systems have provided us with some very important industrial solvents, such as ethanol and acetic acid, for decades. Many surfactants used in chemical manufacturing processes are biological molecules that microorganisms produce naturally, such as emulsan and sophorolipids.

Marine biotechnologists have recently discovered a surfactant produced by marine microorganisms that may replace chemical solvents. However, the biological products that offer us the greatest potential for decreasing the environmental impact of industrial manufacturing processes are the biocatalysts, which are living organisms or simply their enzymes.

Green plastics

Biotechnology also offers us the prospect of replacing petroleumderived polymers with biological polymers derived from grain or agricultural biomass.

Industrial scientists have also genetically modified both plants and microbes to produce polyhydroxybutyrate, a feedstock for producing biodegradable plastics. Finally, biotechnology provides us with the opportunity to produce abundant amounts of natural protein polymers, such as spider silk and adhesives from barnacles, through microbial fermentation. In place of petroleum-based chemicals to create plastics and polyesters, biotechnology uses sugar from plant material. Almost all the giant chemical companies are building partnerships with biotech companies to develop enzymes that can break down plant sugars.

In summary, no matter what stage of industrial production you select-inputs, manufacturing process or final product- biotechnology provides tools, techniques and know-how to move beyond regulatory compliance to proactive pollution prevention and resource conservation strategies that are the hallmarks of industrial sustainability.

Food Processing and Biotechnology Applications

Biotechnology includes a wide range of diverse technologies and they may be applied in each of the different food and agriculture sectors. It includes technologies such as gene modification (manipulation) and transfer; the use of molecular markers; development of recombinant vaccines and DNA-based methods of disease characterisation/diagnosis; in-vitro vegetative propagation of plants; embryo transfer and other reproductive technologies in animals or triploidisation in fish. It also includes a range of technologies used to process the raw food materials produced by the crop, fishery and livestock sectors.

This is the area that will be considered in this conference, the 11th one to be hosted by the FAO Biotechnology Forum since it was launched in March 2000. It is an area that receives relatively little attention from the media, but which is very important for food security in many developing countries. Biotechnology in the food processing sector targets the selection and improvement of microorganisms with the objectives of improving process control, yields and efficiency as well as the quality, safety and consistency of bioprocessed products.

Microorganisms or microbes are generic terms for the group of living organisms which are microscopic in size, and include bacteria, yeasts and moulds. Fermentation is the process of bioconversion of organic substances by microorganisms and/or enzymes (complex proteins) of microbial, plant or animal origin. It is one of the oldest forms of food preservation which is applied globally. Indigenous fermented foods such as bread, cheese and wine, have been prepared and consumed for thousands of years and are strongly linked to culture and tradition, especially in rural households and village communities. It is estimated that fermented foods contribute to about one-third of the diet worldwide.

During fermentation processes, microbial growth and metabolism (the biochemical processes whereby complex substances and food are broken down into simple substances) result in the production of a diversity of metabolites (products of the metabolism of these complex substances). These metabolites include enzymes which are capable of breaking down carbohydrates, proteins and lipids present within the substrate and/or fermentation medium; vitamins; antimicrobial compounds (e.g. bacteriocins and lysozyme); texture-forming agents (e.g. xanthan gum); amino acids; organic acids (e.g. citric acid, lactic acid) and flavour compounds (e.g. esters and aldehydes).

Many of these microbial metabolites (e.g. flavour compounds, amino acids, organic acids, enzymes, xanthan gums, alcohol etc.) are produced at the industrial level in both developed and developing countries for use in food processing applications. A considerable volume of current research both in academia and industry targets the application of microbial biotechnology to improve the production, quality and yields of these metabolites.

Fermentation is globally applied in the preservation of a range of raw agricultural materials (cereals, roots, tubers, fruit and vegetables, milk, meat, fish etc.). Commercially produced fermented foods which are marketed globally include dairy products (cheese, yogurt, fermented milks), sausages and soy sauce. Certain microorganisms associated with fermented foods, in particular strains of the Lactobacillus species, are probiotic i.e. used as live microbial dietary supplements or food ingredients that have a beneficial effect on the host by influencing the composition and/or metabolic activity of the flora of the gastrointestinal tract. Probiotic bacterial strains are also produced and commercially marketed in many developed countries.

In developing countries, fermented foods are produced primarily at the household and village level, where they find wide consumer acceptance. Food fermentations contribute substantially to food safety and food security, particularly in the rural areas of many developing countries. Traditional fermentation processes used in the production of these foods are uncontrolled and are dependent on microorganisms from the environment or the fermentation substrate for initiation of the fermentation processes.

Such processes, therefore, result in products of low yield and variable quality. Microorganisms and metabolic pathways associated with the production of fermented foods are the subject of considerable research, targeting strain isolation and identification; improvement of the efficiency of fermentation processes and the quality, safety and consistency of fermented foods. Much of this research incorporates the use of genetic technologies for strain development and improvement, and for diagnostic studies.

While microorganisms are beneficial in most fermentation processes, some may pose the risk of food contamination and can cause food-borne illness. Diagnostic methodologies which integrate the use of molecular genetic techniques, enhance the speed and sensitivity of microbial testing and are increasingly being applied in developing countries.

Applications of biotechnology to plants or animals to improve their food processing properties (e.g. development of the Flavr Savr tomato variety, genetically modified to reduce its ripening rate) or to produce proteins from genetically modified (GM) microorganisms to improve plant or animal production (e.g. production of bovine somatotropin (BST), a hormone increasing milk production in dairy cows, by GM bacteria) are not considered here. Finally, applications of biotechnology to processing of food and not to processing of non-food agricultural products (e.g. timber) or to applying biotechnology to microorganisms for environmental purposes (bioremediation, biofuels etc.).

Biotechnology in Food Fermentation

Microorganisms are an integral part of the processing system during the production of fermented foods. Microbial cultures can be genetically improved using both traditional and molecular approaches, and improvement of bacteria, yeasts and moulds is the subject of much academic and industrial research. Traits which have been considered for commercial food applications in both developed and developing countries include sensory quality (flavour, aroma, visual appearance, texture and consistency), virus (bacteriophage) resistance in the case of dairy fermentations, and the ability to produce antimicrobial compounds (e.g. bacteriocins, hydrogen peroxide) for the inhibition of undesirable microorganisms.

In many developing countries, the focus is on the degradation or inactivation of natural toxins (e.g. cyanogenic glucosides in cassava), mycotoxins (in cereal fermentations) and anti-nutritional factors (e.g. phytates).

Traditional methods

Traditional methods of genetic improvement such as classical mutagenesis and conjugation have been the basis of industrial starter culture development in bacteria (a culture used to start a food fermentation is known as a starter culture), while hybridisation has been used in the improvement of yeast strains which are widely applied industrially in baking and brewing applications.

Classical mutagenesis: This involves the production of mutants by the exposure of microbial strains to mutagenic chemicals or ultraviolet rays to induce changes in their genomes. Improved strains thus produced are selected on the basis of specific properties such as improved flavour-producing ability or resistance to bacterial viruses. Such mutants may, however, show undesirable secondary mutations which can influence the behaviour of cultures during fermentation.

Conjugation: This is a natural process whereby genetic material is transferred among closely related microbial species as a result of physical contact between the donor and the recipient microorganism. Conjugational gene exchange allows both plasmid-localised and chromosomal gene transfer (a plasmid is a circular self-replicating non-chromosomal DNA molecule found in many bacteria, capable of transfer between bacterial cells of the same species, and occasionally of different species).

Hybridisation: Sexual reproduction in yeasts, and thus genetic recombination, has led to improvements in yeasts. For example, crossing of haploid yeast strains with excellent gassing properties and with good drying properties could yield a novel strain with both good gassing and drying properties.

Molecular approaches

Genetic modification: Recombinant DNA approaches have been used for genetic modification of bacterial, yeast and mould strains to promote expression of desirable genes, to hinder the expression of others, to alter specific genes or to inactivate genes so as to block specific pathways. The successful application of genetic modification for food bioprocessing applications requires the development and use of food grade vectors, i.e. plasmids which do not contain antibiotic resistance genes as markers and which consist of DNA sequences from microorganisms which are generally recognised as safe (GRAS).

GM yeasts appropriate for brewing and baking applications have been approved for use (e.g. approval was granted in the United Kingdom for use of a GM yeast (Saccharomyces cerevisiae) in beer production, containing a transferred gene from the closely related Saccharomyces diastaticus, allowing it to better utilise the carbohydrate present in conventional feedstocks). None of these GM yeasts are, however, used commercially.

Genetic characterisation: The genetic characterisation of microbial strains through the use of molecular diagnostic techniques can contribute tremendously to the understanding of fermentation processes. Molecular diagnostics provide outstanding tools for the detection, identification and characterisation of microbial strains for bioprocessing applications and for the improvement of fermentation processes.

The application of these and other related techniques, along with the development of molecular markers for bacterial strains, greatly facilitates understanding of the ecological interactions of microbial strains, their roles, succession, competition and prevalence in food fermentations and allows the correlation of these features to desirable quality attributes of the final product.

Genomics: In recent years, the genome sequences of many food-related microorganisms have been completed and large numbers of microbial genome sequencing projects are also underway. Functional genomics, a relatively new area of research, aims to determine patterns of gene expression and interaction in the genome, based on the knowledge of extensive or complete genomic sequence of an organism. It can provide an understanding of how microorganisms respond to environmental influences at the genetic level in different situations or ecologies, and should therefore allow adaptation of conditions to improve technological processes.

For a range of microorganisms, it is now possible to observe the expression of many genes simultaneously, even those with unknown biological functions, as they are switched on and off during normal development or while an organism attempts to cope with pathogens or changing environmental conditions. Functional genomics can, for example, shed light on common genetic mechanisms which enable microorganisms to use certain sugars during fermentation, as well as on genetic differences allowing some strains to perform better than others. It holds great potential for defining and modifying elusive metabolic mechanisms used by microorganisms.

Moving from the gene to the protein level, it should also be mentioned that proteomics, an approach aiming to identify and characterise complete sets of protein, and protein-protein interactions in a given species, is also a very active area of research which offers potential for improving fermentation technologies.

Biotechnology in Food Ingredients Production

Flavouring agents, organic acids, food additives and amino acids are all metabolites of microorganisms during fermentation processes. Microbial fermentation processes are therefore commercially exploited for production of these food ingredients. Metabolic engineering, a new approach involving the targeted and purposeful manipulation of the metabolic pathways of an organism, is being widely researched to improve the quality and yields of these food ingredients. It typically involves alteration of cellular activities by the manipulation of the enzymatic, transport and regulatory functions of the cell using recombinant DNA and other genetic techniques. Understanding the metabolic pathways associated with these fermentation processes, and the ability to redirect metabolic pathways, can increase production of these metabolites and lead to production of novel metabolites and a diversified product base.

Many of the classical food microbiological methods used in the past were culture-based, with microorganisms grown on agar plates and detected through biochemical identification. These methods are often tedious, labour-intensive and slow. Genetic based diagnostic and identification systems can greatly enhance the specificity, sensitivity and speed of microbial testing.

Molecular typing methodologies, commonly involving the polymerase chain reaction (PCR), ribotyping (a method to determine homologies and differences between bacteria at the species or sub-species (strain) level, using restriction fragment length polymorphism (RFLP) analysis of ribosomal ribonucleic acids (rRNA) genes) and pulsed-field gel electrophoresis (PFGE, a method of separating large DNA molecules that can be used for typing microbial strains), can be used to characterise and monitor the presence of spoilage flora (microbes causing food to become unfit for eating), normal flora and microflora in foods. Random amplified polymorphic DNA (RAPD) or amplified fragment length polymorphism (AFLP) molecular marker systems can also be used for the comparison of genetic differences between species, subspecies and strains, depending on the reaction conditions used.

The use of combinations of these technologies and other genetic tests allows the characterisation and identification of organisms at the genus, species, sub-species and even strain levels, thereby making it possible to pinpoint sources of food contamination, to trace microorganisms throughout the food chain or to identify the causal agents of foodborne illnesses. Monoclonal and polyclonal antibodies can also be used for diagnostics, e.g. in enzyme-linked immunosorbent assay (ELISA) kits.

Microarrays are biosensors which consist of large numbers of parallel hybrid receptors (DNA, proteins, oligonucleotides). Microarrays are also referred to as biochip, DNA chip, DNA microarray or gene arrays and offer unprecedented opportunities and approaches to diagnostic and detection methods. They can be used for the detection of pathogens, pesticides and toxins and offer considerable potential for facilitating process control, the control of fermentation processes and monitoring the quality and safety of raw materials.

Nutritional Value of Foods

Fermentation processes enhance the nutritional value of foods through the biosynthesis of vitamins, essential amino acids and proteins, through improving protein and fibre digestibility; enhancing micronutrient bioavailability and degrading antinutritional factors. Many bacteria in fermented foods also exhibit functional properties (probiotics). The safety of fermented food products is enhanced through reduction of toxic compounds, such as mycotoxins and cyanogenic glucosides, and production of antimicrobial factors, such as bacteriocins, carbon dioxide, hydrogen peroxide and ethanol, which facilitate inhibition or elimination of food-borne pathogens.

Intellectual Property Rights (IPRs): The processes used in the more advanced areas of agricultural biotechnology tend to be covered by IPRs and these rights tend to be owned by parties in developed countries. This applies also to biotechnology processes used in food processing. On the other hand, many of the traditional fermentation processes applied in developing countries are based on traditional knowledge. In addition to biotechnology processes, microbial strains may also be the object of IPRs. Many of the microorganisms associated with traditional fermentation processes in developing countries are unique.

Biotechnological innovations have greatly assisted in industrialising production of certain indigenous fermented foods. Indonesian tempe and Oriental soy sauce are well known examples of indigenous fermented foods that have been industrialised and marketed globally. The results of biotechnology research will lead to fermented foods of improved quality, safety and consistency.

— Should biotechnology developments in developing countries target commercialisation? Should they target diversification into new value-added products? Should biotechnology development be linked to technological developments in food processing?

— Can the application of biotechnology to food processing allow farmers in developing countries to add value to their agricultural products and improve their revenues?

6

Food Processing through Biotechnology

The word 'food' as used in the food industry therefore has strict meaning - perhaps surprisingly, it also includes drinks, any substance used in the production of food, and even chewing gum. Biotechnology has already benefited the food industry; it has given us high-quality foods that are tasty, nutritious, convenient and safe. As research continues, it seems inevitable that biotechnology will have an increasing impact on the food we eat. It offers huge potential for increasing the range and quality of food available to us, particularly more nutritious, palatable and stable food.

There is an increasing desire among many consumers for food to be as 'natural' as possible, because such food is perceived by consumers to be more nutritious and safer to eat. However, food processing preserves food that would otherwise perish and gives us an interesting and varied choice of products. Most food nowadays undergoes some form of processing.

Traditional biotechnology has played an important role in the development of food supplies for thousands of years. Biotechnology was practised by our ancestors in an effort to find new ways of producing foods. Through trial and error, they learned how to breed and domesticate animals, and developed a variety of food crops. They also developed techniques, such as baking, brewing and fermentation, to process and preserve raw foods. Some traditional biotechnology techniques have been developed much more recently. Nevertheless, such conventional techniques are widely accepted and generally do not cause public concern.

'Modern' biotechnology, in contrast with 'traditional' biotechnology, also uses the techniques of genetic modification—the joining of pieces of genetic material (DNA) in new combinations. This allows characteristics to be transferred between organisms, to give new combinations of genes and improved varieties of plants or microorganisms for use in agriculture and industry. Some consumers are concerned about the safety of using the techniques of modern biotechnology. The techniques of modern biotechnology are becoming an increasingly important part of the overall effort to improve methods of food production and to increase the variety and quality of foods.

Modern biotechnology has potential applications in the production of food, food processing, and also in the assurance of food quality and safety. Although many genetically modified food crops are under development, very few have so far reached the supermarket shelf. The first genetically modified whole food to reach the shops was the FLAVR SAVR™ tomato in the USA, and the first product in the UK was tomato puree.

The recent introduction of genetically modified soybean has caused controversy. Up to 60% of processed foods contain soy or soy derivatives, such as starch, protein or oils. Modern biotechnology is also being used to improve microorganisms, processing aids (e.g. enzymes) or ingredients for use in the production and processing of food. The enzyme chymosin, which is used in the manufacture of 'vegetarian cheese', is an example of this type of genetically modified product.

Need for Food

Food is composed of various chemical elements, arranged in a variety of ways to form molecules. The body uses some of the molecules (nutrients) in food to help it function correctly and stay healthy. There are many different nutrients, which have are variety of functions in the body and are vital to health. Most foods are mixtures of nutrients. No one food provides all of the nutrients our bodies need and in the correct balance, which is why we must eat a variety of foods to meet our nutritional needs.

Like all machines, the human body requires a source of energy. This energy is supplied by food, and is released for use in the body after digestion and absorption. Energy can be supplied by the three main nutrients carbohydrate, fat and protein, although this is not the primary role of protein. Some nutrients, such as vitamins, minerals and trace elements, are very important to health but are required only in tiny amounts - and are thus known as micronutrients. Although not a nutrient, roughage (or dietary fibre) is another important component of our diets. It acts as a laxative, helping to keep our bowels in good working order.

There is general consensus about the type of diet that will maximise health and minimise disease risk. Too much fat in the diet, along with other factors such as smoking and lack of exercise, increases the risk of heart disease. However, some fats contain vitamins or are essential to our health. We also tend to consume too much salt (sodium) and sugar, both of which are often added to our food, either by food processors or ourselves. Too much salt can lead to high blood pressure, which in turn can cause strokes and heart attacks, whereas eating too many sugary foods is a major cause of tooth decay and can lead to obesity (and sometimes diabetes).

Furthermore, sugars contain calories and no other nutrients. Many people consume excess food and therefore energy (calories). The excess is stored as fat in the body, and can lead to obesity. Few of us eat enough fibre and starchy carbohydrates, and possibly not enough of some vitamins and minerals, which are all essential for maintaining good health.

There is increasing evidence for a link between antioxidant nutrients in fruit and vegetables and that these can protect against chronic diseases (including some cancers and heart disease). Eating no fewer than five portions of fruit and vegetables each day, as recommended by the World Health Organisation (WHO), is also an easy way of changing our diet so that we consume sufficient vitamins and minerals, as well as more fibre and starchy carbohydrates. A diet that is high in fruit and vegetables is also likely to contain less fat, sugar or salt.

Improving Diet

Food manufacturers already produce a variety of food products that contain fewer calories (i.e. less energy), and/or reduced amounts of fat, sugar or salt. However, biotechnology offers additional potential to alter the chemical composition of plants to provide us with more nutritious as well as better-tasting and longer-lasting food. A growing number of food products that have a 'healthier' image are available in the supermarket, increasing consumer choice. Thus, we are able to reduce our intake of calories, salt, sugar and/or fats, as well as increase our intake of fibre (e.g. as oatbran), if we so desire.

Fruit and vegetables are good sources of many vitamins. Vitamins are a group of substances that are essential for normal healthy growth and metabolism. Metabolism is the process by which life is maintained and our body converts food into energy. Because some vitamins cannot be made or stored in the body, we must get sufficient and regular supplies from food. Genetic modification could be used to produce fruit and vegetables that contain higher amounts of vitamins, to improve their nutritional quality.

One way to decrease our fat intake is to eat reduced-fat and low-fat products. Biotechnology is being used to develop potato varieties that have a higher starch content than normal. Researchers have introduced a gene from bacteria into potato plants to increase the efficiency of starch manufacture. These potatoes contain 35% more starch, which is the major source of carbohydrate in the human diet. Such potatoes absorb less oil during frying and could be used to produce potato chips and crisps that contain less fat.

Scientists are using biotechnology to alter the content of many oil crops. The degree of saturation of an oil is ultimately determined by the genes in the plant it is produced from. By changing the ability of the plant to make saturated fats, it has been possible to generate different types of oils. Some plant proteins are referred to as 'incomplete' or of 'low biological value' because they lack one or more of the 'essential' amino acids. Attempts are being made to develop soybeans that have a more 'complete' protein composition.

In China, childhood blindness is caused by deficiency in an essential amino acid called lysine. Biotechnology could be used to produce varieties of rice that have improved protein profiles, including higher lysine levels. In many developing countries, malnutrition is due to the lack of protein; thus, for example, genes from pea plants could be transferred to rice plants, to produce varieties of rice that have a higher protein content.

Non-allergenic Foods

Biotechnology could also be used to eliminate allergy-causing proteins from foods, and thus reduce the incidence of food allergy, which can be fatal in some cases. For example, it might be possible to genetically modify peanuts so that they no longer contain the proteins that cause severe allergies. As many as 1 in 200 children in the UK might now be allergic to peanuts. Rice is another potential candidate for genetic modification to remove allergenic proteins, to which some people can be allergic, especially in many Asian countries.

In addition to enhancing the nutritional properties of foods, biotechnology is being used to develop foods that have medicinal properties, so-called functional foods or nutraceuticals. Functional foods are products that are claimed to have a positive health benefit beyond 'normal' nutrition. The benefit is usually due to the addition of a specific bioactive substance, such as the addition of 'biocultures' to 'bio-yoghurts'.

It might be beneficial to increase the levels of the antioxidant nutrients in some foods to improve vitamin-deficient diets, especially in fresh fruit and vegetables. Antioxidants limit or inhibit damage caused by oxidation to molecules, and thus the development of some diseases. Antioxidants that are found naturally in foods help to stop cholesterol picking up oxygen (which is circulating in the blood); thus, cholesterol is deposited less readily in the coronary arteries, reducing the risk of heart disease. Vitamin C in the right amount, may be able to prevent heart disease and cancers: it converts certain cancer-causing compounds (carcinogens) in food to more harmless substances.

The discovery of anti-cancer substances called glucosinolates in vegetables of the Brassica family, which includes Brussels sprouts, cabbage and broccoli, and of the genes coding for them, has prompted research into the development of varieties with enhanced levels of glucosinolates using biotechnology. It is hoped that other vegetables could be modified to produce glucosinolates.

Polyunsaturated fatty acids (PUFAs) are essential to health. Furthermore, foods that are rich in polyunsaturated fatty acids are thought to reduce the incidence of some diseases including heart disease and cancers. Increasing demand for these components (whose principal dietary sources are fish and vegetable oils) has prompted research into the possibility of exploiting fungi for the production of natural sources of polyunsaturated fatty acids, including the omega-3 and omega-6 fatty acids. The fungus Mortierella alpina is already used as a source of polyunsaturated fatty acids, which are added to formulations of baby milk.

Foods with Improved Taste and Qualities

Genetically modified FLAVR SAVR™ tomatoes do not go soft during transport, and can be left on the plant for longer to develop their full flavour. Genetic modification has also been used to slow down softening in other fruit (apple, raspberry and melon), although they are not yet on sale. This benefit is likely to be transferable to other crops including bananas, pineapples, sweet peppers, peaches, nectarines, mangoes and strawberries.

Attempts are also being made to modify some fruit and vegetables, such as tomato, chicory, lettuce and potato, to increase their sweetness and thus make them more palatable to some people.

Slower Ripening Broccoli

The modification of broccoli so that it ripens more slowly, and therefore stays green for longer, is also under development.

There is huge potential for using biotechnology to develop foods with improved processing qualities. Biotechnology is also likely to be used to produce improved microorganisms, both to improve conventional fermentation processes and to develop new ones. These could include microorganisms for the production of foods (e.g. bread, wine, yoghurt and cheese), or for a wide range of fermentation products for use in food processing (e.g. enzymes, vitamins, amino acids and high-grade chemical additives such as citric acid, a flavouring and acidifying agent).

Microbial fermentation Microorganisms, especially bacteria and yeasts, are an important biological tool in the food industry. Many food products and food ingredients are products of microbial fermentation processes. Genetically modified tomato puree is made from tomatoes that have been genetically modified so that they contain high levels of pectin, and thus remain firmer for longer. The tomatoes hold less water because of their high pectin levels, which are retained during processing, and therefore less water has to be removed during processing into puree.

Owing to the improved processing qualities of the genetically modified tomatoes, the resultant puree is claimed to be cheaper to produce. Current research is aimed at using the same genetic modification approach to improve the colour, vitamin content, flavour and storage life of tomatoes, and is being extended to other crops such as bananas.

'Designer' starches: Food processors use starch as a thickener and as a fat substitute in low-fat products. At

present, raw starch is extracted from plants and modified using chemicals or other, mechanical, energy-consuming processes. The modifications required are all very specific and depend on what the starch is to be used for. It might be possible to eliminate some or all of these modification steps, if plants were genetically modified so that they produced starch in a form ready for industrial use.

Genetic modification can be used to produce microorganisms with enhanced properties that are of value to the food industry. Such microorganisms include yeasts that use a wider range of sugars, and bacteria that are resistant to particular bacterial viruses that would otherwise render them useless. Two genetically modified yeasts have been developed and approved for use, although they are not yet in wide commercial use. One raises bread more quickly; the other is for brewing low-carbohydrate beer. New yeast strains that are tailored to the barley and hops that are grown in different regions of the world are being developed for use in the brewing industry.

Enzymes act as biocatalysts in all living organisms. They are already widely used in the food and drink industries, and can be derived from microorganisms, plants or animals. The potential applications of these biological compounds continue to increase as further research and development is carried out. Because they are proteins, enzymes are permanently inactivated by heat, strong acid or alkali, and other conditions that denature (or inactivate) proteins. Thus, enzymes that are required for specific uses are often produced by microorganisms that are selected for their ability to grow under unusual conditions, such as at high temperatures or under acid conditions.

Genetic modification can be used to develop microorganisms that produce enzymes previously obtained from other sources. For example, chymosin (or rennin), which is used to curdle milk in cheesemaking, was previously obtained from calf stomachs. The gene for the enzyme was copied and inserted into yeast cells, which then produce the enzyme. Chymosin is commercially available and used as a rennin substitute in the manufacture of 'vegetarian cheese'.

Many fungi naturally secrete enzymes and have long been exploited commercially as 'factories' to produce useful enzymes for use in food processing. These include enzymes for processing fruit and vegetables, clarifying fruit juice, extracting coffee and spices, and producing sweeteners. Fungi are also used to produce metabolites for use in foods as flavours or preservatives. Genetic modification has been used to improve the enzyme-secreting capacity of some fungi. Research efforts are now being made to engineer fungi to produce large quantities of enzymes that they would not normally produce.

Allergens, and Chemical Residue Contaminants

Nut Allergens

Most nuts, especially peanuts, contain proteins that in some people can cause allergic reactions (e.g. life-threatening anaphylactic shock). People with this allergy must avoid nuts. A test kit that detects the presence of peanuts in foods has recently been developed, for industrial use, with the help of modern biotechnology. In the future, biotechnology might be used to develop nuts that have been genetically modified to remove these proteins, or even vaccines (immunotherapy) to prevent the severe reactions in people with nut allergy.

Similar test kits are also available for the detection of the protein gluten. People who suffer from coeliac disease are sensitive to this protein, which is found in many cereal plants, especially wheat. Gluten damages the lining of the intestines and prevents nutrients from being absorbed. Many commercially available diagnostic kits have also been developed for the detection of residues of pesticides and veterinary drugs.

Microbial Contaminants

It is important to monitor food for spoilage and contamination by microorganisms that can cause serious illness or, at least, make foods taste 'off'. The use of modern biotechnology has proved to be extremely important in the development of rapid and accurate methods for the detection and analysis of bacterial contaminants and pathogens in our foods. Traditionally, the detection of contamination of food with undesirable microorganisms is both labour-intensive and time-consuming. Biotechnology has led to the development of tests that can be completed in a fifth of the time required by conventional test methods.

Fungal toxins

Some fungi produce chemicals (mycotoxins) that are extremely poisonous, such as the cancer-causing toxin aflatoxin, which can be found in contaminated peanuts and corn. Modern biotechnology has led to the development of test kits for the rapid detection of aflatoxin, thus helping to ensure that no affected products are marketed.

Novel Foods for Health: The characteristic smell of broccoli and Brussels sprouts is largely due to a range of chemicals in the plant, called glucosinolates. Because a high level of some of these compounds in the diet might reduce susceptibility to cancers, there is a lot of interest in enhancing the levels of these glucosinolates using biotechnology. Scientists are aiming to increase the levels of glucosinolates in broccoli by introducing genes from wild species.

Brussels sprouts and varieties of broccoli contain chemicals called glucosinolates that deter some insect pests and also give these vegetables their characteristic flavour. Some glucosinolates might provide protection against some forms of cancers, so we should include plenty of these vegetables in our diet.

The yellowing of leaves as they age is due to of an enzyme breaking down the green pigment chlorophyll. The gene for this enzyme has been identified , and when switched off, leaves stay green. Limiting protein breakdown in this way could increase the nutritional value of grasses and other crops, as well as lowering fertiliser costs. A turf species that stays green even after 158 days without water has been developed. This clearly has important uses for ornamental lawns and sports pitches.

7

Foods Derived through Modern Biotechnology

The premarket evaluation is designed to identify endpoints of safety and efficacy, critical to the acceptability of a biotechnology-derived food, and generally negates the need for postmarket monitoring. However, postmarket monitoring may be appropriate under certain conditions where better estimates of dietary exposure and/or nutritional consequence of a biotechnology-derived or other novel food are required.

There may also be a need, in selected instances, to correlate dietary intakes of a nutritionally improved food with expected beneficial effects on human health. There are 2 elements that are essential to the monitoring process. First, adequate data must be available to assess the use, distribution, and fate of the product or commodity within the food supply and second, a monitoring program that provides requisite information on the consumption of relevant foods must be established. The feasibility of identifying any positive or negative effects on human health must be based on a hypothesis with identified biomarkers for measurement and is dependent on the availability of accurate consumption data.

While current exposure assessment principles can be applied to the postmarket monitoring of biotechnology-derived foods to obtain appropriate consumption estimates and confirm premarket predictions, it should be recognised that such assessments may require traceability from field to consumer, which can be very challenging with commodity crops. Exposure assessment methods include both deterministic and probabilistic estimates of intake, using food supply data, individual dietary surveys, household surveys, or total diet studies. If accurate exposure assessments are needed, probabilistic modeling provides a valuable but costly refinement to the coarse evaluation obtained via deterministic approaches.

The postmarket monitoring method used to estimate consumption should be evaluated a priori on a case-bycase basis to determine the most appropriate monitoring tool for establishing a specific relationship between a biotechnology-derived trait and its potential effects on human health.

Premarket Assessment of Foods

The premarket assessment of foods derived from biotechnology is a scientific, risk-based process that ensures the safety of the food to the consuming public. The premarket assessment of foods derived from biotechnology is a step-wise process, with specific safety determinants that vary on a case-by-case basis depending on the nature of the modification and intended use of the food.

In the absence of major compositional changes in the biotechnology-derived food, the premarket assessment strategy focuses on the safety of the identified difference(s). Nutritionally improved foods derived from biotechnology may be intended to have either an enhanced amount or specific composition of major inherent constituents.

Consequently, those biotechnology-derived foods with major compositional changes will require consideration of these changes in the context of existing and expected dietary exposure.

During the premarket safety assessment of any new food, biotechnology-derived or otherwise, it must be established that under the conditions of intended use, there would be no increased safety concern compared to that of traditional or existing foods. Safety can only be evaluated when conditions of use and exposure are known. The resultant exposure estimates are evaluated in the light of data from toxicological and nutritional studies to ensure safety and efficacy.

Postmarket monitoring has not been routine in supporting the safety or regulatory approval of food products, except in unique instances where there has been a need to confirm premarket dietary intake estimates and ensure safety. The call for postmarket monitoring of biotechnology-derived foods with respect to adverse consequences to human health is apparently driven by nebulous concerns for unexpected long-term effects.

Ironically, the notion of required postmarket monitoring for adverse effects from any biotechnology-derived food contradicts the intent of any premarket safety assessment and undermines confidence in any conclusion of safety derived from that assessment. In reality, the identification of a potential food safety risk during the premarket safety assessment of a food from a biotechnology-derived crop would effectively end any endeavor to commercialise that crop. In light of the knowledge of the safety of those foods derived from biotechnology currently on the market, postmarket monitoring requires definition and a strong scientific basis to be a worthwhile endeavor that complements the premarket assessment.

Moreover, there is no basis to conclude that postmarket monitoring should be applied only to biotechnology-derived foods. It is not evident that inherent differences exist between conventionally derived foods and those derived from biotechnology that would necessitate the use of postmarket monitoring specifically in one context but not the other.

Postmarket Monitoring of Foods

Postmarket monitoring is not standard practice in supporting the safety of a food product. However, the regulatory approval of certain novel foods and food additives has been contingent upon the conduct of monitoring programs. It is notable that these studies were conducted in the absence of regulatory guidelines on the design or conduct of postmarket monitoring for foods.

Passive surveillance as " the collection of voluntary responses by consumers without input from the producer, distributor, or regulator." Passive surveillance or monitoring may be useful in cases where immediate idiosyncratic adverse effects, such as allergenicity, are of concern. However, the scope of the hypotheses that can be answered via passive monitoring and, therefore, its utility without further follow-up is limited. The confirmation of any health effect attributed to the consumption of a food requires the use of appropriately validated diagnostic tools.

Active monitoring may be appropriate when confirmation of premarket consumption projections is required, when effects identified through passive monitoring require targeted investigation, or when longterm monitoring for potential chronic health effects is of interest. In the latter case, it is arguable that postmarket monitoring no longer constitutes monitoring per se, but falls instead into the realm of epidemiology, and that appropriate epidemiological techniques should therefore be applied.

Recent mandates within the EU have made environmental postmarket monitoring of GM organisms obligatory, and a preliminary guidance document for notifiers and risk assessors has been published. The directive requires that manufacturers design and implement an environmental postmarket monitoring plan with the aim of gathering information on the immediate and cumulative long-term effects of exposure to GM organisms. The mandate includes both general monitoring for unanticipated adverse effects and, where necessary, the hypothesisdriven confirmation of potential effects identified through the premarket environmental risk assessment process.

In addition to the premarket safety assessment required by the European Novel Food Regulation, postmarket monitoring of GM foods and feeds and products derived from GM food ingredients has also been proposed. By definition, the proposed directive requires the implementation of both passive and active measures, with specific initiatives targeted on a case-by-case basis as a result of the risk assessment process.

Any postmarket monitoring program will be of limited value unless precise estimates of exposure to a particular product are available. Consequently, the collection of adequate consumption data and methods for ensuring the traceability of the biotechnology-derived food of interest are crucial elements in this process. The premarket safety assessment of biotechnology-derived foods is designed to identify areas of safety or nutritional concern before the introduction of these foods for general consumption.

Therefore, postmarket monitoring programs may be appropriate when specific hypotheses requiring directed investigation have been identified as a result of the premarket assessment and the data generated from premarket studies are unable to address the particular hypothesis in question. Ill-defined use of postmarket monitoring is unwarranted, and could be considered misguided, as it implies that postmarket monitoring is an appropriate means of hazard identification, as opposed to a control option for risk management possibly undermining the reliability of or the confidence in traditional approaches for determining food safety.

Due to the resource intensive nature of postmarket monitoring, definite goals for a monitoring program should be established a priori, particularly when attempting to determine whether postmarket monitoring is necessary and/or appropriate.

Applications of Postmarket Monitoring

The investigation of idiosyncratic adverse events or chronic health effects, the confirmation of premarket exposure estimates, or the identification of changes in dietary intake patterns, represent examples, where in certain instances, hypotheses may be appropriately tested through postmarket monitoring programs. The success of any postmarket monitoring strategy is dependent on the accurate estimation of exposure in targeted or affected population groups; the ability to measure a specific outcome of interest, either directly or through appropriate biomarkers; and the control of confounding factors.

Idiosyncratic Adverse Events

The potential allergenicity of foods derived from biotechnology has been recognised as relevant for particular consideration because some of the modifications in these foods may include the expression of proteins not otherwise present in that food. The presence of proteins per se does not merit attention as a concern regarding the safety of foods derived from biotechnology. However, the current definition of food allergens (for example, proteins eliciting an IgE-mediated hypersensitivity response in sensitive individuals following consumption) dictates that an assessment for allergenic potential is in order when the biotechnology-derived food or food ingredient contains a novel protein that would otherwise not be present in food.

A premarket decision tree strategy for assessing the allergenicity of GM foods has been recommended by the Food and Agriculture Organisation of the United Nations and the World Health Organisation and has proceeded to Step 8 of the Codex Alimentarius Commission's assessment of the safety of foods derived through biotechnology. The lack of a single definitive assay predictive of the allergenic potential of a food protein limits this strategy as a positive predictor of allergenicity. However, an absence of positive results in the overall outcome of the decision tree provides reasonable assurance that the likelihood of allergenicity is low.

Therefore, based on the scientific assessment applied prior to the commercialisation of a biotechnology-derived food, a postmarket monitoring program implemented strictly to address allergenicity should not be necessary. Food allergies represent a unique challenge in ensuring the safety of any food introduced into the population, biotechnologyderived or otherwise.

Implementation of a postmarket monitoring program for allergenicity strictly to complement premarket assessment data, or appease acceptance of scientific uncertainty, should be tempered by the knowledge that allergic reactions to food represent individualistic responses in that the majority of the population can consume the offending food without adverse reaction. It would be reasonable to expect that the entire population would not be susceptible to allergic reaction following consumption of a biotechnology-derived food. Caution must be exercised in the monitoring approach taken and efforts should be made to distinguish those populations that may be at particular risk, as defined by the specific conditions under which postmarket monitoring is attempted.

Guidance toward susceptible populations of individuals should be available from the multiple assessment tests conducted in the premarket evaluation. A postmarket monitoring program to assess the potential allergenicity of a food derived from biotechnology would represent a passive reporting program for spontaneous adverse events occurring under conditions of acute postmarket exposure. In addition, the program must contend with elements related to exposure assessment (for example, accurate estimates of consumption, traceability, and so on) in order to provide monitoring data of use.

Furthermore, such a program must be aware of further challenges that exist due to limitations in the extent of current understanding regarding food allergens. While it is reported that greater than 90% of all food allergies worldwide are caused by proteins associated with the consumption of cow's milk, crustaceans, eggs, fish, peanuts, soybeans, tree nuts, and wheat, or ingredients derived from these foods, the properties of a protein that make it allergenic, as well as the biological variables governing a person's susceptibility to developing an allergy, remain to be determined.

Allergic reactions to food clearly represent a special case of food safety. Research is ongoing; however, the minimum amount of allergenic protein and duration of exposure necessary to induce sensitisation are not yet known. Additionally, once an individual is sensitised, the level of subsequent exposure to the allergenic protein needed to trigger an allergic response requires delineation.

Health Effects

The association between diet and health is well recognised. Notwithstanding this fact, it is difficult to establish a causal relationship between a particular dietary component and a specific health endpoint, even when an extensive amount of research has been conducted. The use of a postmarket monitoring program to identify chronic adverse health effects not otherwise indicated by the premarket safety assessment would be highly impractical, if not scientifically impossible. This would require the prospective long-term monitoring of a specific population in the absence of a hypothetical outcome, or alternatively, the retrospective identification of potential exposure to a particular biotechnology-derived food in a group of individuals known to have a specific disease or health condition.

Apart from the special case of allergenicity, which could be considered an acute adverse event, the potential for adverse effects based on the chronic consumption of biotechnology-derived foods would be no different than those associated with the chronic consumption of other foods, and by definition should be minimal, because the approval of a food is dependent on its premarket assessment of safety. A food with a potentially beneficial effect on human health, as indicated in premarket assessment studies, may require confirmation of its purported benefits through postmarket monitoring or conclusive demonstration of a positive diet-health relationship in clinical trials.

With the advent of biotechnology-derived foods designed to provide specific nutritional benefits or improved functional characteristics, the use of postmarket monitoring to detect physiologically relevant beneficial health effects may be more appropriate and more readily applicable than the situation depicted for adverse effects. In this case, postmarket monitoring would constitute a traditional targeted epidemiological study specifically designed to monitor the long-term benefits of a particular biotechnology-derived food.

Difficulties associated with implementing this type of investigation would be similar to the difficulties associated with any long-term dietary study, particularly in terms of compliance, mitigating or confounding factors, contributing risk factors, and the inherent heterogeneity of the study population in terms of predisposition to effects. Continual changes in both food availability and individual dietary habits also make it difficult to establish a direct causal relationship between any observed health benefits and the long-term consumption of foods derived from biotechnology.

A key step in the premarket safety assessment process is the generation of dietary intake estimates and their comparison to published guidelines, such as acceptable daily intakes for food additives, tolerable upper intake levels for nutrients, or provisional tolerable weekly intakes for contaminants. A tiered approach to exposure assessment is generally advocated, whereby crude screening tools are used to dictate the need for more highly refined consumption estimates.

The premarket exposure assessment may provide an initial screening of consumption, as it is often based on several conservative or "worst-case" assumptions that result in overestimates of intake. When no toxicologically significant exposure is indicated as a result of this assessment, further refinement would not be warranted. In certain situations, postmarket monitoring may provide valuable confirmation of premarket consumption projections. For instance, the premarket exposure assessment may not capture realworld exposure scenarios, or a more accurate estimate of intake may be desired for comparison with upper limits of safety.

As well, premarket exposure estimates may not adequately characterise the consumption of specific population groups or certain age groups. In these situations, postmarket assessment provides an additional measure of safety, through the validation of premarket assumptions regarding anticipated intake levels and additional sampling to determine patterns of exposure in specific subgroups of the population, particularly those individuals at potentially greater risk.

Changes in Dietary intake Patterns

With the advent of foods derived from GM crops with improved nutritional characteristics or improved functionality, consumers may actively choose products based on nutritional or physiological benefits. Consequently, actual levels of exposure and any subsequent impact on dietary consumption patterns, and therefore nutritional status, may not be evident until after the market introduction of the biotechnology-derived food product.

Alterations in the micro-or macronutrient profile of a biotechnologyderived food could significantly affect dietary intakes, causing nutrient deficits or, alternatively, causing intakes to approach or exceed the upper level. Similarly, nutritional status may be affected if the biotechnology-derived food product alters consumption patterns via the displacement of other core foods or food components.

In specific cases, these considerations may necessitate the use of postmarket monitoring. However, the presence of confounding factors makes the longterm monitoring of consumption patterns difficult and minimises any observed associations between diet and health, because those individuals who make a nutritionally significant or healthy food-choice are likely to also have significantly different dietary patterns from individuals who do not make similar choices.

Population Exposure

Tracking the disappearance of biotechnology-derived foods into the food supply. The implementation of a postmarket monitoring strategy designed to identify positive or negative effects on human health is dependent upon the availability of accurate consumption data. Exposure assessment methods that have traditionally been used to quantify the intakes of low-molecular weight chemicals, micro- and macronutrients, and whole foods, could theoretically be applied to the postmarket monitoring of biotechnology-derived foods to obtain appropriate consumption estimates and to confirm premarket predictions.

Exposure to a particular substance can be quantified if the food chemical concentration and consumption of a particular food are known or can be estimated with a certain degree of reliability. There are 2 elements in this process. Adequate data must be available to assess the fate of the product or commodity within the food supply. Secondly, a monitoring program that provides requisite information on the consumption of relevant foods must be established. The hypotheses of the monitoring program dictate the level of accuracy required in the generation of intake estimates via an exposure assessment, which in turn are dependent on the precision of the supporting data.

Therefore, the contribution of exposure assessment to the overall postmarket monitoring process is directly related to the quality of information underlying the evaluation. This requirement emphasises the need for setting study objectives a priori, to ensure that adequate and appropriate data are collected for subsequent analyses and that potential future uses of the data are anticipated. In the event that existing data are used to estimate the intake of novel foods or food components, such as biotechnology-derived foods or foods containing biotechnologyderived ingredients, this also highlights the need for standardised methods of data collection in terms of both food consumption and food composition, to enable meaningful comparisons of results among different countries.

To appropriately quantify the concentration of a biotechnologyderived food or ingredient therefrom at the level of the processed food material or brand, processes for ensuring the traceability of the commodity or food ingredient must be readily available to document the movement of such material through the feed and food chains. If particular food products containing biotechnology-derived ingredients could not be identified with a high level of accuracy throughout all levels of food production, processing, and distribution, the monitoring of latent health effects from biotechnology-derived foods would be very difficult, if not impossible.

For some, there is a tendency to consider postmarket monitoring not on a case-by-case basis, but as the complete representative monitoring of adverse health effects from exposure to foods derived from biotechnology. For this scenario to generate useful data, a central repository for the biotechnology content of specific foods or primary agricultural commodities would need to be available and current food composition databases would need to be expanded to incorporate industry-specific information.

In addition, if the aim of a postmarket monitoring program was to monitor potential long-term effects on human health, temporal information on food composition would need to be included in these databases. Historical, current, and future levels of biotechnology-derived food components in the food supply would need to be accurately measured and predicted in order to limit discontinuity in food composition information. Clearly, this would be a time-intensive and not particularly cost-effective effort, especially in view of the low risk already assured by the premarket assessment process.

Integrating disappearance data with food consumption databases. The accuracy of the exposure assessment for foods derived from biotechnology or ingredients therefrom is dependent on the model used and the validity of the underlying parameters. To assess the intake of specific food components, mathematical methods for integrating food composition and consumption data include, in ascending order of refinement, point estimates, simple distributions, and probabilistic modeling or Monte Carlo analysis.

Point estimates combine a single estimate of food chemical concentration with a corresponding fixed, generally "worst-case," estimate of food consumption. Alternatively, simple distributions combine the distribution of exposures with a fixed value for chemical concentration. Lastly, probabilistic analyses include a measure of the variability associated with both food consumption and food composition data. By utilising data from these respective distributions, an estimate of the probability of various exposure outcomes can be calculated through iterative mathematical modeling.

Probabilistic modeling is especially relevant when considering the introduction of nutritionally improved foods derived from biotechnology, where several different products may be modified for a similar nutritional benefit or physiological health effect. Monte Carlo assessment could determine the probability of aggregate exposure to several different biotechnology-derived food sources with the same functional characteristics and therefore compare intended food uses with consequent exposure in particular target groups, susceptible subgroups, or the population in general.

Consequently, depending on the data collected, the specific endpoint being investigated, and the precision required from the exposure assessment, an appropriate method of analysis can be selected. When considering the long-term health effects of foods derived from biotechnology, habitual or lifetime intake estimates may be necessary. However, the overestimation of either food consumption or the biotechnology-derived content of relevant foods in a long-term epidemiologic study would lead to the overestimation of exposure. Therefore, the risks or benefits associated with the biotechnology-derived food would subsequently be underestimated.

Similarly, if an acute endpoint such as allergenicity is of interest, in which a single critical exposure may precipitate an adverse event, reliable data must be available regarding the consumption of relevant foods, because knowledge of the intake required to induce or elicit a response is limited. If accurate exposure assessments are needed, probabilistic modeling provides a valuable refinement to the coarse evaluation obtained via deterministic approaches. However, the cost and resources needed to implement these methods vary substantially, and the potential to determine causation may be limited.

The postmarket monitoring method used to estimate consumption should be evaluated a priori on a case-by-case basis to determine the most appropriate monitoring tool for establishing a specific relationship between the nutritional enhancement or characteristic trait of the food derived from biotechnology and its potential effects on human health.

Demonstration of Causality

One intended outcome of a postmarket monitoring strategy for a food with some purported effect on human health may be the demonstration of causality. The consideration of a valid scientific hypothesis supported by measurable parameters is a required element of a postmarket monitoring program intended to distinguish a cause-effect relationship between the consumption of food and a specific health effect to the consumer.

In the absence of defined hypotheses, data mining to investigate health-related effects is likely to identify random correlations not limited to the consumption of foods derived from biotechnology. While the safety of foods derived from biotechnology is established in the premarket assessment, and efficacy may be demonstrated in pre- or postmarket studies, the ability to discern unpredicted adverse effects, or intended beneficial effects on health is a challenge given the myriad of confounding variables encountered under real-world conditions.

Correlations that have not been adjusted for confounding factors may be spurious and not causally related to dietary exposure to a food derived from biotechnology. Certain confounding or contributing risk factors should be anticipated before implementing a postmarket monitoring program, and these risk factors should be minimised through proper statistical design and analyses. Furthermore, the ability of any data collected in a postmarket monitoring program to demonstrate causality may be limited by the characteristics of the population included in the study program.

In a postmarket monitoring program, establishing that reported effects on human health are the result of exposure to a consumed food cannot rely on the notion of definitive proof, but rather on the weight of evidence to establish a scientific basis of association.

8

Environmental and Food Analysis

Antibodies "made to measure" are required for both single and simultaneous measurement of multiple analytes. Chemical pesticides, polychlorinated biphenyls (PCBs) and antibiotics are typical examples of contaminants which can be rapidly and efficiently determined by antibody based methods. Antibodies enable not only a rapid detection of analyte in water, body fluids, soil or food extracts, but they can also be exploited in sample preparation prior to analyte detection. Because antibody affinity and specificity determine primarily the analytical capability of the immunochemical method, the properties of the antibodies represent an important innovative factor in developing an analytical system.

Highly sensitive detection of toxic analytes can be performed by enzyme immunoassays, immunosensors and related techniques while immunoaffinity chromatography and flow injection immunoassay systems enable the concentration and clean up of the analytes in question. Low-molecular weight compounds are called haptens in immunology. Preparation of antibodies against haptens such as pesticides and PCBs is based on covalent binding of the hapten to a carrier protein and immunisation of animals by the synthesised immunogens.

The manner of chemical binding of the hapten to a protein determines the character of the antibody specificity. Various types of hapten derivatives conjugated to proteins have been used for antibody development. Polyclonal antibodies are produced by using traditional immunisation procedures, namely in rabbits, goats, sheep and pigs. A certain drawback of the method lies in the fact that it is not possible to produce identical antibody specificity even in two animals of the same species.

On the other hand, wide spreading of monoclonal antibodies is limited by low predictability of the hybridoma technology result. It can be noted that polyclonal (rabbit) antibodies are the most widely used reagents in immunochemical analysis so far. However antibody engineering and production of recombinant antibodies is a very promising field both for research and application. Basic synthetic ways of preparation of the hapten derivatives (hapten design) have been explored primarily in steroids in the eighties of the last century.

Hapten immunochemistry thus represents a consistent area for the development and preparation of conventional antibodies. As new impulses for further progress are largely depleted, further experimental strategies have been sought outside the classical immunochemistry area. In the last decade, molecular biology has generated fundamental changes in antibody production. Preparation of recombinant antibody fragments with novel binding properties was a primary goal of gene technologies. Their major asset lies in the possibility to focus mutagenesis to that part of gene which determines the structure and affinity of antibody binding site.

Large phage libraries expressing antibody fragments on the surface of individual phage particles were used for preparation of recombinant antibodies. The systems enable separation of individual phage particles and subsequent selection of phage antibodies from a large number of expressed phage particles.

Immunochemical Strategies

Phenoxyacetic Acid Herbicides

The effort was focused on 2,4-dichlorophenoxyacetic acid (2,4-D) and 2-methyl-4-chlorophenoxyacetic acid (MCPA) representing the group of broadly used phenoxyalkanoic acid herbicides which can potentially contaminate ground water supplies. 2,4-D and MCPA molecules were conjugated to protein via their carboxyl group by the conventional mixed anhydride chemistry, forming the desired herbicide-protein conjugates.

The conjugates served as immunogens for the immunization of animals or as coating antigens in indirect ELISA, polarisationfluoroimmunoassay or immunosensors. A panel of specific monoclonal antibodies against 2,4-D was produced in this laboratory by various clones. Indirect and direct assays employing these antibodies were optimized in respect of coating dilution, enzyme conjugate (tracer), pH, reagent volumes and incubation times to reach maximum assay sensitivity.

The sensitivity of direct monoclonal ELISA was about 0.1 µg/l. The piezoelectric immunosensor based on these monoclonal antibodies exhibited a limit of detection below 0.01 µg/l. Monoclonal antibody (clone E2/G2) was used for construction of an afinity biosensor based on solid supported lipid membranes. The antibody fragments showed a lower degree of specificity and stability but their practical utilisation as immunoassay reagents is limited at present. The antibodies produced against MCPAprotein immunogens exhibited always high crossreactivity with 2,4-D herbicide.

s-Triazine Herbicides

Atrazine and simazine are the most common monochloro-substituted derivatives belonging to the s-triazine family. Owing to high persistence in the environment, accumulation of s-triazines in soil may cause ground water contamination. Moreover, atrazine carry-over may be harmful to certain rotation crops and therefore the soil should be analysed before planting.

The hapten-protein immunogens were prepared by derivatizing atrazine and simazine and subsequent conjugation of the derivatives to proteins via aminohexane carboxylic and thiopropane carboxylic groups. The rabbit antibody produced allowed the determination of atrazine in water samples in the range of 0.02 to 5 μg/l in direct ELISA format. A polarisation fluoroimmunoassay for atrazine was optimised by structure changes of labelled antigen.

Monoclonal antibodies raised against the same hapten-protein immunogens were produced and subsequently employed for development of piezoelectric and amperometric immunosensor. The monoclonal reagents incorporated into the immunoassay system were less sensitive compared with rabbit antibodies. ELISA methods using both rabbit and monoclonal antibodies were used for determination of atrazine in methanolic extracts of real field samples. Analysis of the soil samples by ELISA showed good correlation of results obtained by immunoassays and gas chromatography within the concentration range of 0-380 μg/kg.

The data obtained by rabbit and monoclonal ELISA showed no difference although the sensitivity of the method using rabbit antibody was one order of magnitude higher than that based on monoclonal antibody. Additional monoclonal antibody, having a generic pattern for simazine and related s-triazines, was employed for the development of a highly sensitive flow through amperometric immunosensor based on the peroxidase chip and enzyme-channelling principle.

Preparation of modified disposable screen-printed electrodes using enzymechannelling can be a promising approach when the detection of extremely low analyte concentration is required. Monoclonal antibodies against atrazine, simazine and 2,4-D were used for the development of the flow injection immunoanalysis and compared with conventional immunoassays. ELISA was the more reliable technique in real screening analysis.

Sulfonylurea Herbicides

The growing popularity of the sulfonylurea herbicides in the agriculture industry has necessitated analytical methods for water and soil matrices with good recovery and low detection limits. A single ring hapten strategy was suggested to produce generic antibodies with dominant selectivity towards arylsulfonyl or triazine moieties of metsulfuronmethyl herbicide. Superior monoclonal antibody showed in indirect ELISA cross-reactivity among eight related sulfonylurea species and proved to be an effective generic reagent for this class of herbicides.

The assay operated within the ppt-ppb calibration range. A decline in assay sensitivity was observed in monoclonal antibody batches originating from later production stages. It should be noted that such change in properties has not been observed in other monoclonal antibodies produced in this laboratory.

Polychlorinated Biphenyls

PCB are recognized as ubiquitous environmental pollutants, because of their low rate of environ-mental degradation and tendency to accumulate in tissues. They represent a class of 209 discrete congeners with 1-10 chlorine atoms attached to a biphenyl nucleus. In the Veterinary Research Institute, a considerable effort has been devoted to produce analytical antibodies against PCBs within the last 15 years.

Goat antibodies, incorporated into radioimmunoassay, were applied to the screening of PCB in cow's milk. A correlation of results of immunoassay and gas chromatography ($r = 0.67$, $n = 27$) was found within the concentration range of 0.1-5.58 mg Aroclor 1260/kg milk fat. Sheep antibodies raised against low-chlorinated PCB congeners were produced and subsequently used for screening of PCB in soil extracts by ELISA.

The antibodies were further used for a development of disposable electrochemical immunosensor for the detection of PCB in food samples. Antibodies against highly toxic (coplanar) PCBs and respective enzyme tracers were developed in recent years. The established direct ELISAs were highly sensitive (ppt-ppb level) and specific for non-ortho (coplanar) PCB 77, 126 and 169 congeners whereas assay response to more abundant but less toxic PCB congeners was very low or negligible. The antibodies in combination with alkaline phosphatase tracers are being employed at the University of Florence for development of electrochemical enzyme immunoassay for coplanar PCBs. Commercial ELISA kits for coplanar congeners were developed in collaboration with Abraxis LLC (USA).

Surfactants and Nonylphenol

Due to the extreme amounts of surfactants emitted into sewage plants and the persistence of some of their known metabolites, such as nonylphenol, this is a group of environmental pollutants with the highest priority. Although legislators prescribe biodegradability of surfactants, these cannot be completely mineralized in biological waste treatment plants, thus they arrive together with their metabolites in rivers and some are found even in drinking water. Rabbit antibodies against linear alkyl benzene sulfonates (LAS) have been produced and incorporated into direct ELISA format or polarisationfluoroimmunoassay format.

The direct ELISA detection limit for LAS (90% technical mixture of 4-dodecylbenzenesulfonic acid isomers) was about 2 µg/l. It appeared from the cross-reactivity results that the developed ELISA does not distinguish parent LAS homologues and isomers from their major carboxylic metabolites. Thus, measurement of real samples by the ELISA will provide screening values reflecting a total amount of LAS including their carboxylic derivatives. Polyclonal and monoclonal antibodies were prepared to establish direct assays for determination of nonylphenol and octylphenol. 4-nonylphenols are toxic metabolites on non-ionic synthetic detergens having oestrogenic effects and resistance towards biodegradation.

Rabbit antibodies were employed for establishment of assays with sensitivity about 3 µg/l. Specificity of monoclonal antibody (clone 4H6) allowed a sensitive detection of the long-chain forms of 4-n-alkylphenols. No cross-reactivity interference was indicated for linear alkylbenzensulfonates and phenolic compounds. These antibodies were employed for development of ELISA and dipstick tests to measure nonylphenol in water samples and for development of polarisationfluoroimmunoassay. A rapid Biacore biosensor immunoassay of 4-nonylphenol was optimised and applied for shell-fish analysis.

Veterinary Drugs

The nitrofuran antibiotics were banned from use in food animal production in 1993-1995 in Europe and the Czech Republic because of concerns about their carcinogenicity and mutagenicity. Effective control of nitrofuran drugs (furazolidone, nitrofurantoin, furaltadone, nitrofurazone) can be achieved by measuring tissue concentrations of bound nitrofuran residues (3-amino-2-oxazolidinone (AOZ), 1-amino-hydantoin (AHD), 3-amino-5-methylmorpholino-2-oxazolidinone (AMOZ) and semicarbazide (SEM)) which are toxic metabolites of their parent compounds. Within the European project FoodBRAND, monoclonal antibodies for detection of AOZ, the metabolite of furazolidone, were produced and characterised in terms of specificity and sensitivity (Vass et al., unpublished yet).

The superior antibodies, produced by clones 3B8/2B9 and 2D11/A4, were exploited for development of the simplified ELISA method without using extraction and clean up operations. Detection capability of the method was 0.4 μg/kg for shrimp, poultry, beef and pork meat. An excellent correlation of results obtained by the ELISA and LC/MSMS within the concentration range 0-32 μg/kg was found in the naturally contaminated shrimp samples and in the incurred poultry samples. A testing ELISA kit for AOZ was developed in collaboration with the Queen's University of Belfast and R-Biopharm, Darmstadt. An effort of this laboratory has been devoted to producing antibodies against sulfonamides.

A highly sensitive immunoassay for determination of sulfadimidine in tissues was established. However, a need to detect various sulfonamide species within a single assay prompted us to develop antibodies for a broad-specificity test. At present, a panel of classspecific (generic) antibodies have been produced. Polyclonal antibodies with a broad specificity for simultaneous detection of 18 sulfonamides at the maximum residue level have been prepared and characterised in both direct and indirect ELISA formats (unpublished results).

Immunoassays in Food Research

The trends in research can be followed by analysis of the published papers related to the specific topic used as a key word for searching in a suitable database. Web of Science database offers not only the widest selection of journals indexed but the software tool for such analysis as well. Immunoassays, namely in radioimmunoassay format (RIA), opened the new era in endocrinology in the seventies of the last century.

In a certain sense, ELISA and biosensors seem to have assumed a similar importance in environmental and food science since the nineties. The increase of applications of ELISA is obvious from an index, ELISA/RIA, which is 0.65 and 1.87 in 1980-1990 and 1991-2004, respectively. The increase in the number of publications on RIA, ELISA and biosensors is 4.2, 12.1 and 28.3 times, respectively, in the same periods 1980-1990 and 1991-2004 (Table 1). These numbers of publications are clear evidence of the general importance of antibody based methods in recent decades.

Table 1. Number of publications

	1980–1990	1991–2004	Index
Radioimmunoassays	5 860	24 639	4.2
ELISA	3 821	46 152	12.1
Biosensors	420	11 871	28.3

The antibody based methods' significance for environmental and food analysis is evident from the data resulting from the analysis of published papers. In the nineties the number of publications dealing with ELISA or immunoassays and biosensors in food and environmental analysis dramatically increased.

9

Food Processing and Food Consumption

Food production is by far the dominant form of land use. The human diet varies widely in different parts of the world, but Figure 1 illustrates the composition of the average food intake in 1996. Some items, such as alcoholic beverages derived from cereals and fruits, are excluded. Cereals remain the mainstay of the human food supply, but significant changes have taken place over the past four decades. Per capita consumption of cereals and meat rose 17 percent and 36 percent respectively, while per capita consumption of fish and seafood rose 57 percent. Consumption of leguminous plants such as rice and soybeans, valued for their high protein content, rose in absolute and per capita terms.

Vegetables, starchy roots and pulses all declined in importance on a per capita basis, though total consumption increased. Also significant has been the increase in consumption of what might be termed "lifestyle" products such as coffee, tea and alcoholic beverages. Coffee production, for example, has risen by 37 percent since 1961 and the harvested area now accounts for about one percent of total arable land. Rising incomes, together with the development of sophisticated food processing and distribution systems and a liberalised world trade regime,

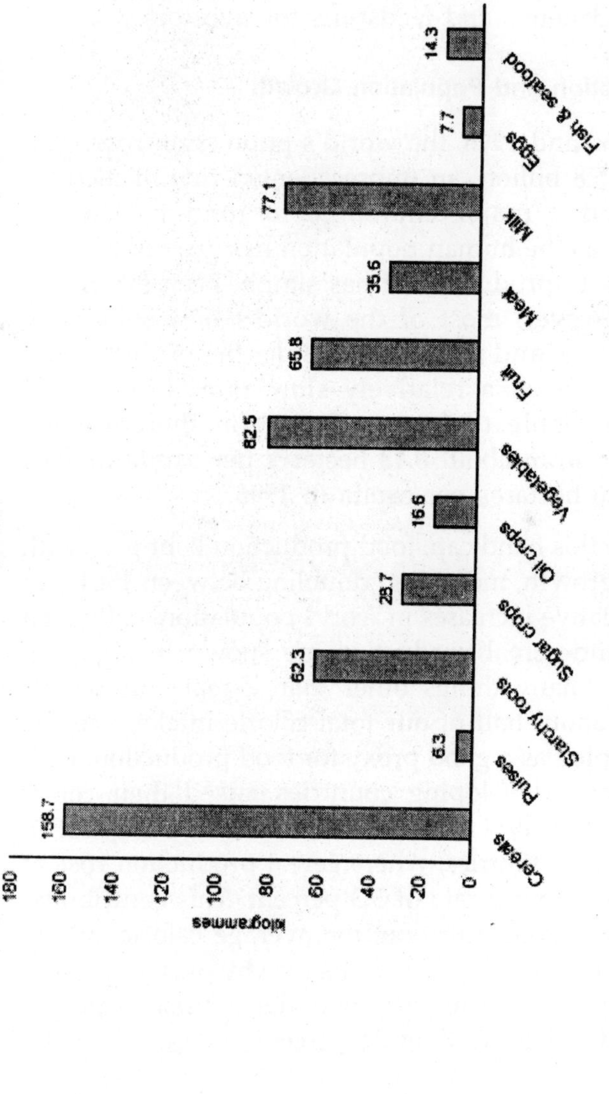

Figure 1. World Food Consumption, 1996 (kg/capita/year)

have enabled consumers in the industrialised countries, and many affluent centers in the developing world, to enjoy unprecedented choice in the foods they eat. As a result, many countries have re-oriented part of their agriculture sector towards export products including "exotic" food items, and feedstuffs for livestock.

Food Production and Population Growth

Between 1960 and 1996, the world's population rose from 3 billion to 5.8 billion, an unprecedented rate of increase which posed a major challenge to food producers. Historically, as the human population has grown, the area of land used to produce food has simply been expanded. By 1960, however, most of the world's best soils were already in use and new land could be brought into cultivation only at a relatively slow rate. The area of cropland available to feed each person therefore fell dramatically, from about 0.43 hectares per capita in 1961 to about 0.26 hectares per capita in 1996.

Despite this handicap, food production kept pace with population growth, more than doubling between 1961 and 1996. The relative increases in world population, cultivated land area, and cereal production are shown in Figure 2. Humans eat many things other than cereal, but cereals account for about half of our total calorie intake, and are usually accepted as a good proxy for food production. Both developed and developing countries raised their cereal production after 1961, but growth rates were highest in the developing countries, where cereal production rose at an annual compound rate of 3.3 percent. This contributed greatly to improved nutrition; the average calorie supply to people in less developed regions rose by over 30 percent and the proportion of hungry or under-nourished people in the world fell from 35 to 21 percent.

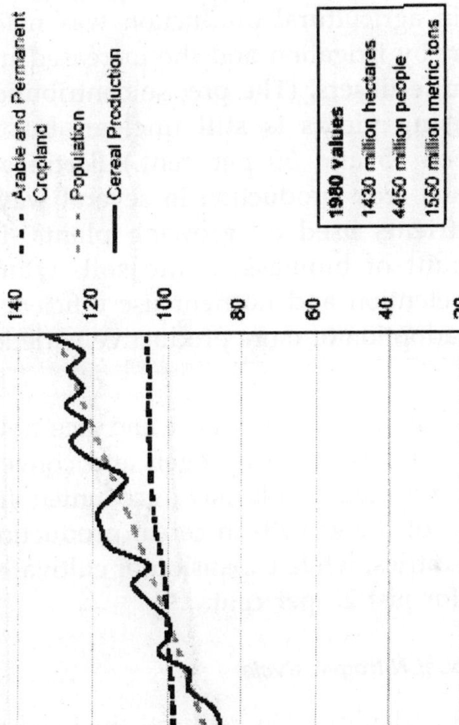

Figure 2. Population, Cultivated Land, and Cereal Production, 1961-1996, *Indexed to 1980*

Nevertheless, because of population increase, the absolute number of undernourished people fell only slightly, and now stands at about 840 million people. The world's revolution in agricultural production was made possible in large part by irrigation and the increased use of inorganic nitrogen fertilisers. (The precise contribution of fertilisers to raising yields is still unclear, but is estimated at between 35 and 50 per cent.) Fertilisers contribute to increased crop production in several ways: by replenishing nutrients used by growing plants, by increasing the amount of biomass in the soil, which improves moisture retention and nutrient use efficiency, and by enabling the adoption of more productive varieties of cereal.

New, high-yielding varieties of wheat and rice have been bred specially to utilise more nitrogen and convert it into more grain. Increased yields have accounted for more than 80 percent of the growth in cereal production in the developing countries, while expansion of cultivated land has accounted for just 20 per cent.

Agriculture and the Global Nitrogen Cycle

The intensification of agriculture has reduced the need to expand cultivation into marginal and ecologically fragile areas but there have been other, considerable, environmental costs. Poor soil management, over-grasing, and inappropriate use of pesticides and herbicides are associated in many parts of the world with soil erosion, desertification, fertility declines, and contamination of soils and water supplies.

Firstly, modern agriculture is the leading source of anthropogenic (human origin) nitrogen entering the environment. Inorganic nitrogen fertilisers are an essential input to maintaining high crop yields. They cannot readily be substituted and fertilisation must probably increase in the future if we are to feed a still-growing world population.

In addition, meat consumption is rising world-wide and the numbers of livestock, and associated volumes of nitrogen-rich manure, will rise too. Secondly, in the past decade, we have greatly advanced our understanding of the global nitrogen cycle, and it has become clear that human addition of nitrogen to the environment is disrupting entire ecosystems across a wide geographical range. What have usually been considered as distinct problems -for example, eutrophication of surface waters, or acidification of lakes and forest soils - should be seen as symptoms of a more universal assault on the global environment.

The global scale of nitrogen pollution is, at present, under-appreciated. Policy responses to the challenge of meeting future food needs without further unbalancing biological systems remain under-developed. There is a need for more longterm and coordinated thinking, and the closest analogy lies with the emergence of international efforts to manage the world's energy system and excessive emissions of carbon. Disruption of the global nitrogen cycle now appears to warrant the same degree of attention.

Global consumption of fertiliser has risen spectacularly, increasing tenfold between 1950 and 1989 (Figure 3). The steep drop in consumption after that date was due principally to collapsing demand in the former Soviet Union, and Central and Eastern Europe, but there was also a substantial fall in Western Europe, caused by grain surpluses, low crop prices, and saturated markets.

The increasing share of nitrogen in the global fertiliser mix becomes clear after about 1960, and the bias is most strongly pronounced in the developing countries, where

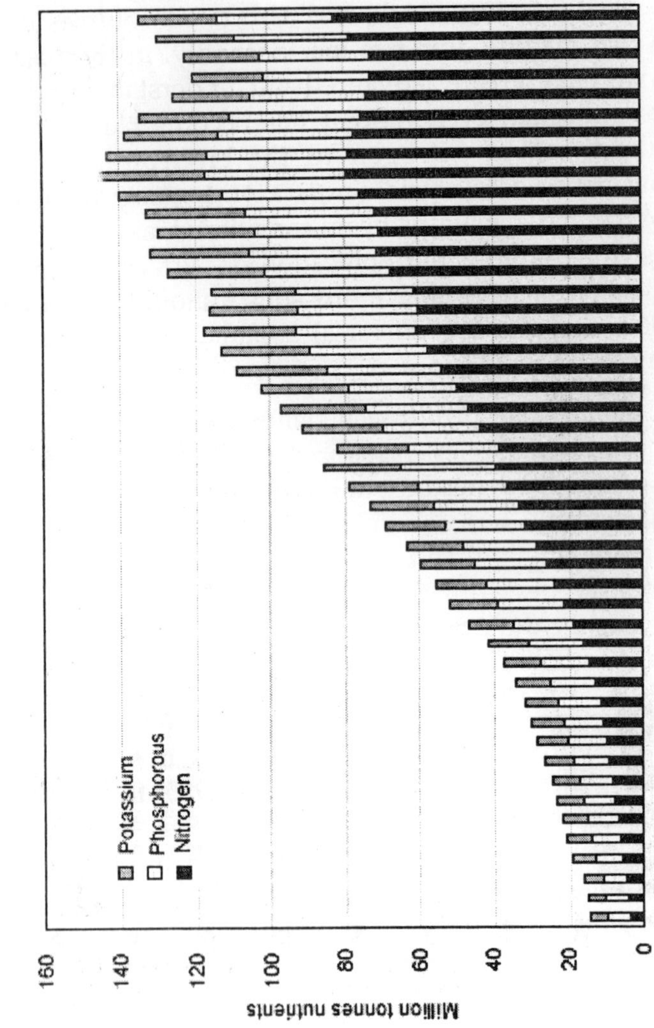

Figure 3. Global Fertilizer Consumption, 1950/51-1996/97

nitrogen now accounts for 66 percent of fertilisers consumed, compared with 55 percent in the developed countries. The bias towards nitrogen fertilisers has been encouraged in part by increased production in areas where cheap natural gas is available and in part by the perception that nitrogen delivers the most spectacular yield gains, at least initially.

The pattern of global fertiliser consumption has changed markedly over the past 30 to 40 years (Figure 4). In 1960, the developing countries accounted for just 12 percent of all consumption; today the figure is nearly 60 percent. Fertiliser use in the developing world has been fuelled by rapid population growth and growing demand for food grains. This is especially true of Asia, where the scope for land expansion is limited.

By contrast, the industrialised countries (with the exception of the former Soviet Union) increased their fertiliser consumption only marginally after 1980; population growth was low, most people were adequately fed, and world agricultural exports had stagnated due to economic problems in many importing countries. Asia is now the dominant player, accounting for 50 percent of world fertiliser consumption, and 86 percent of developing country consumption. Fertiliser application rates vary widely among the major world regions and, perhaps surprisingly, are not strongly correlated either with national income or with need (as indicated by low soil fertility or food insecurity). Fertiliser use varies from a low of 10 kg/hectare in sub-Saharan Africa to a high of about 216 kg/hectare in East Asia (by nutrient weight).

The world average application rate is about 83 kg/hectare; the developing countries as a whole just exceed this figure, and the developed countries as a whole fall just short of it. Fertiliser consumption has been strongly promoted in the industrialised countries and in Asia by

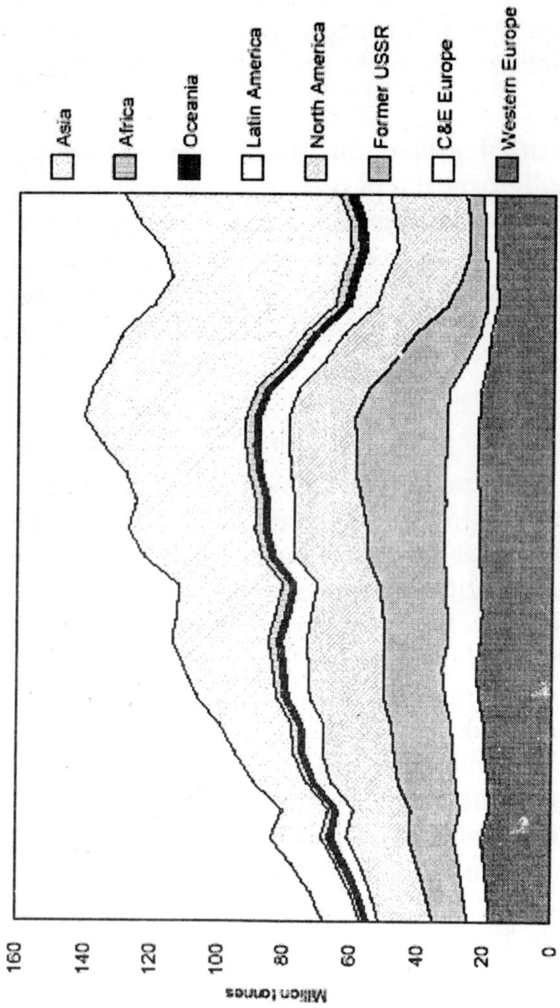

Figure 4. Fertilizer Nutrient Consumption (NPK), 1970-1996

investment in production capacity and a policy and fiscal framework encouraging liberal fertiliser use. Latin America has experienced wide fluctuations and modest overall growth, due to an unstable policy environment and economic crises throughout the 1980s.

Production and consumption in sub-Saharan Africa remain very low, despite the urgent need to raise yields. Growth has been hindered by foreign exchange shortages, low crop prices, policy instability, and inadequate physical and institutional infrastructure. More than half the countries in the region are wholly dependent on fertiliser aid to meet their requirements.

A second trend of key importance of food consumption and the nitrogen cycle is the growing popularity of meat and dairy products in the human diet. With rising income, consumers choose to eat more meat; still greater affluence and concerns for a healthy lifestyle seem to encourage a shift from red meat to poultry. Meat production world-wide has tripled since 1961, reaching 213 million tonnes in 1997 (Figure 5), with output gains concentrated in the United States, the European Union, and China. Individual consumption remains highest in the industrialised world.

Average per capita meat consumption in the United States was 118 kg/year in 1996, while the average for the developed world as a whole was 76 kg/person/ year. Average per capita consumption in the developing countries was 24 kg/year but the picture is rapidly changing in many parts of South America and Asia. For example, the Chinese each consumed 41 kg of meat in 1996, up from 20 kg/ year just a decade earlier.

Total meat consumption in the developing countries just exceeds that in the developed world and, in internationally traded meat and meat products, there is a small net inflow to the developing countries. To meet

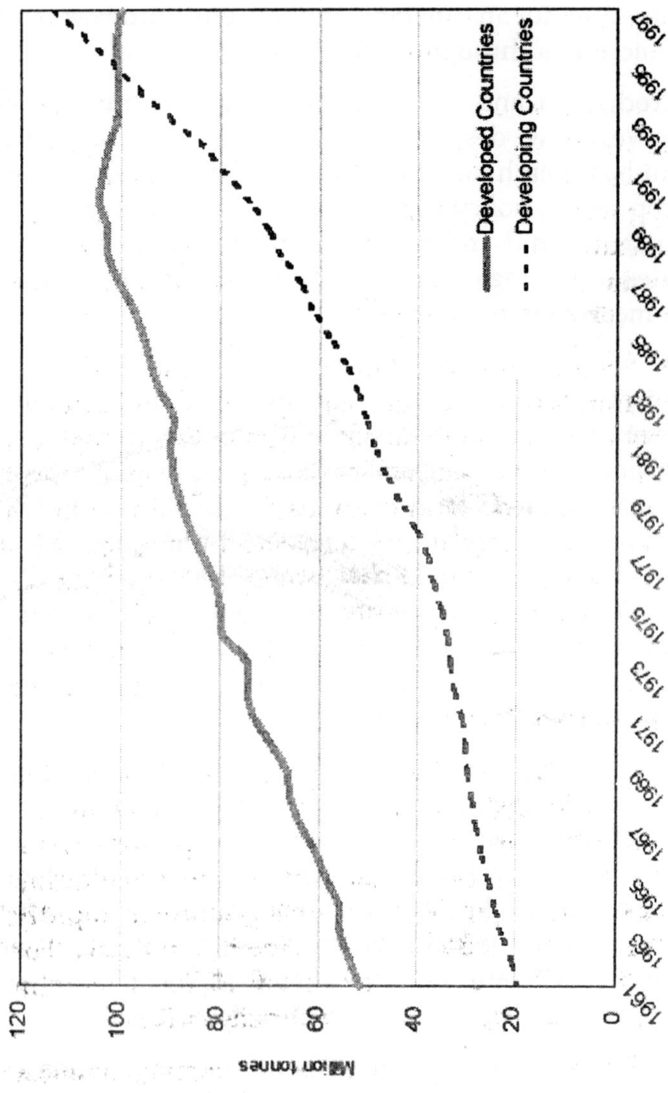

Figure 5. World Meat Production, 1961-97

growing demand, the world's livestock population has boomed. Cattle numbers rose by 40 percent between 1961 and 1997, pigs by 130 percent and chickens by 246 percent. The world today is home to 13.5 billion chickens.

In industrialised countries, animals traditionally reared on rangelands or in farmyards are now increasingly concentrated in intensive feedlots, where they are fed on cereals and commercial preparations of grain, animal protein, and fish meal. This trend, in turn, is leading to the concentration of huge volumes of manure, which cannot economically be redistributed back to areas where the cereals were originally grown. The nitrogen component of manure varies according to the animal and its diet but a crude global estimate is that approximately 32 million tonnes of nitrogen (derived from fodder crops and forage) are deposited into the environment via manure each year.

Where animals are still free-ranging, manure may act as a fertiliser. Where animals are concentrated in feedlots, manure is increasingly viewed as a waste disposal problem. Data for the United States indicate that, of nearly 160 million tonnes of manure produced annually, some 60 percent is excreted directly onto pasture and cropland, while 40 percent is collected from animals in confinement and must somehow be disposed of.

Manure volumes have reached a critical point in parts of Northwestern Europe, where nitrogen deposition far exceeds the absorptive capacity of crops. Concentration of livestock in feedlots is not yet the norm in developing countries, though industrialscale chicken farms are becoming more common, for example, in some South American countries.

Fertilising the Earth

Great uncertainties are involved in measuring the global distribution and transport of nitrogen, how much is being stored, and where, because reactive nitrogen in its many forms is highly mobile, moving easily between terrestrial, freshwater and marine ecosystems, and the atmosphere (Figure 6). But enough is known to be certain that human domination of the nitrogen cycle is responsible for serious pollution and disruption of biological processes which underpin—among other important functions—food production.

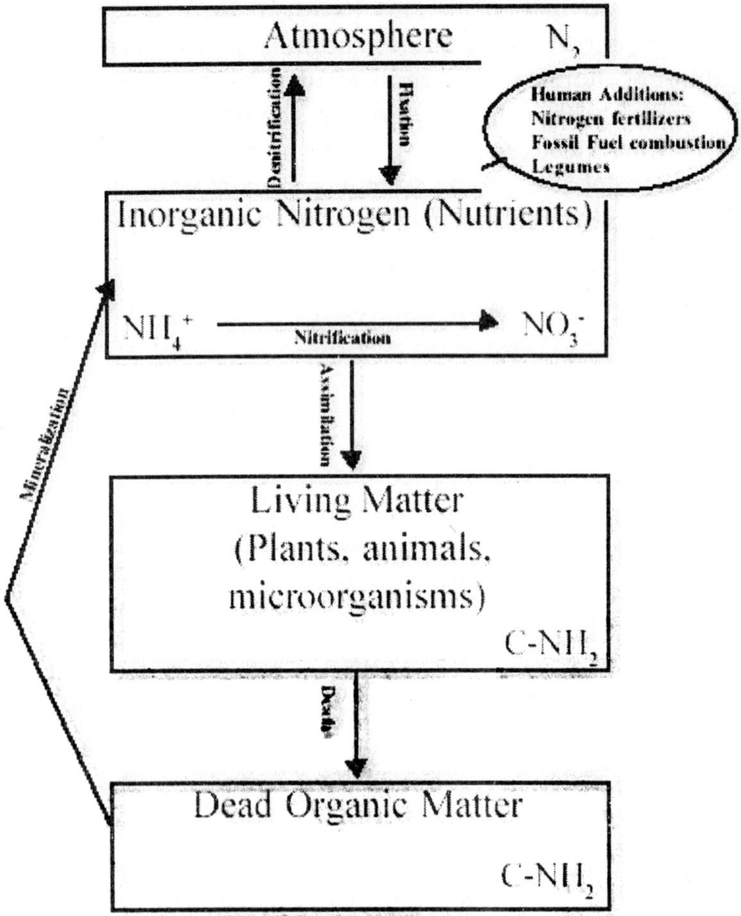

Figure 6. Simplified Version of the Terrestrial Nitrogen Cycle

Human activity is now fixing nitrogen (creating reactive nitrogen from nonreactive N in the atmosphere) at least as fast as natural terrestrial processes. To some extent (which varies according to ecosystem characteristics) additional nitrogen can be utilised by plants and their productivity will be enhanced. Beyond that point, excess nitrogen will accumulate in the environment, changing the chemistry of soils and often reducing their fertility, displacing grassland species, suffocating aquatic plants and fish, and touching off toxic algal blooms.

Increased deposition of nitrogen into the environment has two principal effects. Oxidised forms of nitrogen (NOx and nitrates) acidify natural ecosystems and generally degrade them. For example, scientists now report that acid rain leaches forest soils of up to 50 percent of their calcium, potassium, and magnesium—crucial minerals which buffer or neutralise acids and are essential for plant growth. When soil chemistry is changed dramatically in this way, it is likely to take many decades for all the linked ecosystems to recover.

A related potential impact is toxification. Leached soils acidify and lose their ability to bind and sequester heavy metals, these metals are thus released for uptake by plants. Much of the "forest death" which struck Germany and other Central European countries in the 1970s may have been caused by aluminum poisoning. Toxics accumulated over decades may continue to be mobilised and leached into ground and surface waters, or taken up by plants and into the food chain.

The same applies in acidified lakes, where aluminum and other metallic ions which were formerly bound to soil particles can be mobilised. In contrast, reduced forms of nitrogen (ammonium ions) fertilise whole ecosystems, increasing their net primary productivity but also disrupting them. Some forests in Europe are reported to be growing much faster in the second part of this century than they did in the first half. The disturbing explanation appears to be that trees are absorbing nitrogen directly from the air into their leaves and bark. It is estimated that, in northern Europe, such above-ground uptake now accounts for 60 percent of the nitrogen found in broadleaved trees.

While the extra nitrogen increases growth, trees cannot regulate their intake as they can with nitrogen taken up via their roots; acidification of the soil also tends to leave them weak and vulnerable to insects and mildew. Another consequence of over-fertilisation is loss of species diversity, as nitrogen-responsive plants crowd out others with lower tolerance.

In the Netherlands, intensive livestock production and high human population density have generated the highest nitrogen deposition rates in the world; the conversion of species-rich heathlands and pastures to species-poor grasslands and forests is well documented. Similar trends have been reproduced on experimental grasslands in both the United Kingdom and the United States. However, there is a limit to the amount of nitrogen which plants can absorb.

At some point, other nutrients become the limiting factor, no more nitrogen can be taken up, and additional deposits are simply dispersed into surface or ground water and the atmosphere. This state is known as nitrogen saturation; it has already been reached across large areas of northern Europe, and to a lesser extent in northeastern USA. It is now known that a substantial portion of atmospheric nitrogen deposited in the former region moves directly from air, to land, to water, without ever being taken up by any living organisms or playing any role in biological systems.

Freshwater Ecosystems

Fertiliser is the dominant source of nitrogen washed into the world's water courses, though domestic sewage is another important source (the human diet is rich in protein). Deposition from the atmosphere is another major source in some parts of the world. An ironic aspect of intensive fertilisation is that, on average, only half the additional nitrogen is taken up by plants, due to saturation, or local weather conditions, or because nitrogen is applied at an inappropriate point in the plants' growing cycle.

As a result, nitrogen runs off the soil. It is either carried out in eroded soil, or leached out in the form of dissolved nitrate, carrying with it positively-charged minerals such as potassium and calcium. Much of the nitrogen is later denitrified by bacteria, that is, "unfixed" back into the atmosphere, while other nutrients are washed away. The net result is that nutrients can be "mined" from the soil, leaving it more impoverished than before.

Nitrogen run-off has doubled since pre-industrial times, which represents both a loss of nutrients and an economic loss to farmers. Nitrogen concentrations in freshwater, so far, have been studied most intensively in industrialised countries. For example, a 1994 national survey in the United States revealed that nearly 40 percent of the country's lakes and rivers were too polluted for basic uses such as fishing or swimming; the leading source of pollution in both categories was agricultural run-off. However, nitrogen is fast accumulating in developing countries too.

In China, nitrate concentrations in the Yangtze River (Chiang Jiang) increased fourfold between 1963 and 1980, while concentrations of ammonium approximately doubled. Much of this increase is attributable to the use of inorganic nitrogen fertilisers, which rose by a factor of about 12 over the period. Nitrogen run-off contributes to the phenomenon of eutrophication of freshwater lakes and rivers - essentially fertilisation in the wrong place. Elevated loads of nutrients, chiefly nitrogen and phosphorous, stimulate abundant growth of algae and other aquatic plants.

When the extra plant matter dies, it sinks to lower depths and decays, depleting the water's supply of dissolved oxygen, and killing other aquatic organisms such as deepdwelling fish. The result is scum-covered, malodorous water, unfit for drinking or recreational activities. In many freshwater systems, phosphorous, not nitrogen, is the limiting growth factor. However, phosphorous is also supplied by agricultural fertilisers, as well as by industrial and household effluents.

Eutrophication is increasingly well-documented world-wide, and is of particular concern where freshwater fish are an important source of local food. According to the FAO, some 18 percent of the global annual fish catch comes from freshwater sources. Nitrate contamination of drinking water supplies is widely recognised as a serious health threat. If exposed to high nitrate levels, young infants may develop the potentially fatal "blue-baby" syndrome, where their red blood cells cannot function properly, and fail to deliver sufficient oxygen.

Adults risk contracting a variety of cancers, although the risks remain unclear. Levels of nitrates in drinking water have been closely monitored for many years in the industrialised countries, and the data confirm a historic rise in nitrogen levels in surface waters. In major rivers of the northeastern United States, nitrate concentrations have risen three- to tenfold since the beginning of the century.

Nitrate contamination is also the USA's most widespread groundwater pollution problem; in a national survey, 22 percent of wells in US agricultural areas contained nitrate levels in excess of the federal limit. Some communities in the worst affected States now provide free bottled drinking water to at-risk households (those with infants under six months and pregnant women) when local nitrate levels rise too high, and many are beginning to install costly purification systems.

Nitrates are a prime contaminant in Europe, where the European Commission has enacted a Directive establishing maximum permitted levels in drinking water and requiring costly purification or mitigation measures. In 1,000 lakes in Norway, nitrate levels doubled in less than a decade. Nitrate pollution is also widespread in Australia, and in parts of Africa, Asia, and the Middle East. No global assessment has been undertaken of nitrate pollution, but individual

Marine Ecosystems

Eutrophication of estuarine and coastal zones has emerged as an immense and growing problem in recent years. Over 40 million tonnes of nitrogen, in dissolved and particulate form, are transported by the world's rivers into estuaries and coastal waters each year—double the pre-industrial rate. Unlike freshwater systems, where phosphorous is usually the limiting growth factor, nitrogen is usually the limiting growth factor in saline waters. Additional nitrogen can therefore promote huge algal blooms and significant oxygen depletion (hypoxia) in lower-depth waters.

Some of the best-documented examples of coastal eutrophication come from the United States. According to a recent survey, 52 percent of the nation's estuaries suffer from some degree of oxygen depletion. The worst affected area is the Gulf of Mexico, where 85 percent of estuaries are affected. In the most dramatic example, a so-called "dead zone" of 16,000-18,000 km has developed where the

Mississippi River discharges into the Gulf. Fish and shrimp have disappeared from the area, threatening the local fishing industry, while less mobile life-forms, such as starfish and clams, have died.

Scientists have linked the growth of the dead zone to nitrogen fertilisers and livestock manure from farms hundreds of miles upstream. More than half the 11 million tonnes of nitrogen added to the Mississippi Basin annually comes from fertiliser, and only about 50 percent is taken up by plants. Nearly 2 million tonnes of nitrogen flow down the Mississippi each year, more than triple the amount 40 years ago and the dead zone has ebbed and flowed consistently with peak river discharges and the associated nutrient flux.

The Mississippi dead zone is one of more than 50 similar oxygen-starved coastal regions which now exist world-wide, a threefold increase over the past 30 years. Since coastal zones are among the world's richest fishing grounds, unchecked agricultural run-off poses a serious threat to commercially important fish stocks. Nitrogen pollution is blamed in part for the collapse of the Baltic Sea cod fisheries in the early 1990s, as well as major fish kills (and associated human illness) following outbreaks of Pfiesteria, such as that affecting the Chesapeake Bay, in the United States, in the summer of 1997. Toxic algal blooms, known as "red tides" or "brown tides" are growing world-wide in frequency and severity, damaging offshore fisheries and causing losses to aquaculture enterprises.

The world's population is projected to grow to about 7.3 billion by 2020, with over 90 percent of the increase occurring in developing countries. More people will eat more food, and more protein since, as already described, people almost invariably choose diets which are richer in meat and dairy products as their incomes rise. This will require more cereal to be grown per capita: nearly 40 percent of total grain production is already fed to livestock and the grain-to-protein conversion efficiency is low, lying between 2:1 and 7:1.

The International Food Policy Research Institute has recently projected that global demand for cereals will increase by 41 percent between 1993 and 2020, and that meat demand will rise 63 percent to 306 million tonnes. Most projections of agricultural output assume that demand for food will be met, at least at the global level, though regional and local shortages are likely to persist.

A scenario developed by the Stockholm Environment Institute forecasts cereal production of just over 3 billion tonnes in 2025, while the FAO has produced a shorter-term production forecast of 2.3 billion tonnes in 2010. Given current production of just under 2 billion tonnes, these and other estimates indicate that production increases of up to 50 percent will be required over the next 25-30 years. The scope for expansion of the agricultural land base is limited. Potentially fertile land still exists to be opened up for cultivation, but gains will be partially offset by the loss of productive land to soil erosion, degradation, and urban development.

The FAO suggests that an additional 90 million hectares of land might be brought into production by 2010, mostly in South America and Africa, which amounts to a modest increase of barely 6 percent on today's cropland area. This alone will not be enough to offset the present trends under which both available cultivated land area and grain harvest on a per capita basis are declining sharply. In 1984, the grain harvest per person peaked at 346 kg; by 1995 it had fallen to 295 kg, the lowest level since 1967.

Given circumstances of a rising population, a growing appetite for meat and dairy foods, a static or declining cropland base and declining per capita cereal production, the imperative of increasing yields becomes clear. Agricultural science has, to date, relied on three means of achieving higher yields: genetic improvements of crops, advances in agronomic practices, and synergies between the two. As an example of genetic improvement, plants have been bred to raise the share of photosynthetic product going to the seed, rather than to leaves, stems or roots.

Today's wheat, rice, and corn have a "harvest index" of about 50 per cent, compared with 20 percent before improvement. Further gains appear limited, since scientists believe the theoretical maximum for photosynthetic redistribution to the seed is about 60 per cent. Agronomic practices have focused on increased levels of external inputs, chiefly fertilisers, but also pest and weed control agents, and water, delivered by irrigation.

Synergies have occurred, for example, new dwarf varieties of cereal, with short stalks, are able to utilise two or three times the "normal" amount of nitrogen and develop rich, heavy heads of grain without collapsing. This synergy explains the doubling and tripling of yields achieved during the first phases of the Green Revolution. However, scientists have not found it possible to maintain this momentum, and the rate of yield growth in all major cereals has either plateaued or declined over the past decade. It is often argued that increased fertiliser application in the developing world will overcome yield stagnation and raise harvests closer to the theoretical maximum.

More fertiliser will certainly be part of the answer, but inherent environmental constraints, especially water scarcity, as well as infrastructural and institutional shortcomings, are likely to prove a greater obstacle to the world-wide advent of "industrial" farming, with its dependence on external inputs.

Analysts at the International Fertiliser Development Center (IFDC) have made detailed projections of future fertiliser demand, and come to a problematic conclusion. Their "real world" projection takes into account various economic and non-economic variables, such as foreign exchange availability, crop and fertiliser prices, the development of irrigation and other infrastructure, and the impact of policy reforms. On this basis, global consumption of fertiliser is projected to rise from 134 million tonnes today to 208 million tonnes in 2020 (Figure 7).

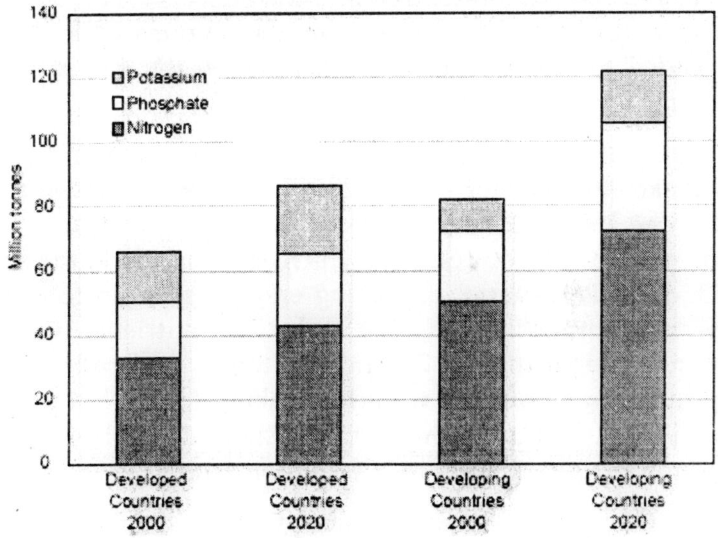

Figure 7. Fertiliser Demand Projections, 2000 and 2020

The strongest growth is expected in Africa (2.4 percent annually), Latin America (2.2 percent annually), Asia (1.8 percent annually) and Central and Eastern Europe (2.2 percent annually). These rates of increase are far below those experienced between 1960 and 1990, reflecting a higher baseline, and changed economic and policy environments. Nitrogen consumption would rise from 82 million tonnes to 115 million tonnes, an increase of about 40 per cent. This projection is comparable with another recent estimate of 134 million tonnes, which assumes higher growth rates in Africa and Latin America. This substantial increase might not be enough, however.

The IFDC's second projection is based on food production needs, that is, the amount of fertiliser needed to grow 2.5 billion tonnes of cereal or more. This leads to global fertiliser consumption in 2020 of 263 million tonnes, and nitrogen consumption of some 160 million tonnes. The fertiliser "shortfall" in developing countries might be as much as 64 million tonnes. A third projection is based on "sustainable farming" needs, or the amount of fertiliser needed to maintain nutrient reserves in the soil at their initial levels.

The projection is based on assumptions about nutrient uptake efficiency for various crops, expected improvements by the year 2020 in different world regions and the proportion of crop residues which are returned to the soil. Under this scenario, producing enough grain to feed the world, without mining the soil of its nutrients, would require application of 366 million tonnes of fertiliser; the proportion of nitrogen would probably decrease but would still be substantial. The fertiliser shortfall in developing countries is estimated at 130 million tonnes.

Globally, fertiliser consumption must grow, but it must grow in a far more managed and thoughtful way than it has over the past 40 years. The objective of agriculture in coming decades must be to optimise soil productivity while preserving its capacity to function as a healthy, complex ecosystem. Fertiliser consumption will remain essential to food security in coming decades, but disruption of terrestrial and marine ecosystems could be greatly reduced if greater efforts are made to control the flows of nitrogen through the environment.

On the production side, inorganic fertilisers should be priced and regulated more in accordance with their environmental impacts (removal of perverse incentives) and used more efficiently (innovative and integrated farming practices, new crops, nitrogen recycling). On the consumption side, the choices we make about diet and the sources of food we buy could, over time, help to reorient the world's food industry to a less nitrogen-intensive base.

Nitrogen Management on the Farm

The efficiency of nitrogen fertiliser use today is very low. The proportion of nitrogen taken up by plants varies widely with crop, climate conditions, and agricultural practices, but field trials at experimental stations indicate plant uptake levels of 50 to 70 percent during the fertiliser application season. On farms, nitrogen losses can be much greater.

Numerous studies indicate that nitrogen loss rates of between 20 and 50 percent are still common in the industrialised countries, while nitrogen uptake by crops in Indian and Chinese rice paddies has been measured at 25 to 30 percent, principally because of rapid losses to run-off, erosion or gaseous emissions. Such wastage represent a substantial economic loss, as well as an environmental hazard. Given that over 80 million tonnes of nitrogen were applied to fields in 1996, a (conservative) 50 percent loss at a wholesale price of US$0.66 per kg of nitrogen in urea, amounts to US$26.4 billion.

Improved on-farm practices can significantly reduce nitrogen pollution. Fertiliser use efficiency has been improving in some developed countries, where farmers are being encouraged to plant cover crops in winter, avoid poor cropping practices, and plant "buffer" vegetation between fields and water courses to trap nitrogen run-off.

Further improvements might be expected from more widespread adoption of slow-release fertilisers and urease inhibitors which can reduce the leaching of nitrate and/ or emissions of nitrous oxide and losses of ammonia through volatilisation. Currently, use of these products is inhibited by low prices for conventional fertilisers.

Synthetic slow-release fertilisers are inherently more expensive to manufacture and only about half a million tonnes are applied annually world-wide, mostly to high-value crops and in non-agricultural sectors such as golf courses and gardens. Fiscal instruments which reduce the price differential between slow-release and conventional fertilisers could speed market penetration by the more efficient product. In the longer-term, a significant contribution to improved nitrogen management would seem possible with new crop varieties and changes in the global balance of crops cultivated. Bio-engineered crops such as wheat and corn which can fix their own nitrogen could reduce the need for fertiliser application.

More research is needed to understand the potential benefits and drawbacks of cultivating leguminous plants such as alfalfa and soybeans. Currently, legume planting world-wide adds about 40 million tonnes to the total amount of nitrogen fixed annually. Legumes are "free" natural fertilisers and their introduction into crop rotations, and ploughing of residues back into the soil, if more widespread, could assist in maintaining soil structure and fertility.

Integrated farming practices, which seek to maximise soil fertility and stable yield gains through crop diversity and adaptability to local conditions will also have a role to play in reducing dependence on high levels of fertiliser application. In many developing countries, fertiliser application levels at present are too low to cause significant eutrophication or acidification damage and a different policy approach is required. However, while nitrogen saturation might seem a far distant prospect in most of Africa and South America, tropical soils are fragile - often thin and light - and nitrogen is more readily lost to run-off or volatilisation to the atmosphere.

The potential for losses and environmental damage is therefore high and the agronomically optimum level of application may be lower than in temperate regions. A priority area of attention should be the correction of unbalanced fertilisation. Most fertilisers used in the developing world are nitrogenbased, because of their low cost per unit of nutrient and the quick and evident response of the plant.

However, increased yields deplete the soil of other major and micro-nutrients which are removed with the harvested crop. These nutrients become deficient unless they are replenished. Excessive nitrogen applications, relative to potassium and phosphates, can lead to nutrient "mining" and loss of soil fertility. In many countries, especially in sub-Saharan Africa, nutrient removal already exceeds nutrient replacement by a factor of three or four. Yields can also be reduced due to "lodging" (where the crop is unable to support its own weight), and greater competition from weeds and pest attacks.

In much of Asia, rice yield losses of between 20 and 50 percent have been recorded as a result of disproportionate applications of nitrogen. Significant yield increases could thus be achieved through the encouragement of inorganic fertilisers which provide a more balanced mix of nitrogen, phosphates, and potassium. In many developing areas, however, lack of access to capital, lack of roads and rail links which could ensure timely delivery of fertilisers, and inadequate information, are likely to mean that soils continue to be gravely under-fertilised.

The concept of managing a global nitrogen cycle helps to illuminate the fact that intensive agriculture and high protein diets are tending to concentrate nitrogen in some parts of the world, while impoverished farmers, without fertilisers, burning their crop residues and animal wastes for fuel, see their soil nutrients disappearing. The trend is accelerated by the growth of international trade in food, especially grains, which redistributes nitrogen from net producer to net consumer countries.

Economic development and poverty alleviation are the best long-term redress. In the shorter-term, internationally-funded research efforts into alternative methods of plant nutrition and support for traditional agricultural practices which help to maintain soil would be well repaid in terms of enhanced food security, reduced financial losses, and avoided environmental damage.

Complementary measures include greater use of organic fertilisers (animal manure, human wastes, crop residues), applications of lime on the acid soils which are typical of tropical regions, alternating the growth of shallow-rooted and deep-rooted plants to hold nitrogen in the soil, and the use of crop rotations incorporating leguminous plants. Integrated plant nutrition systems (IPNSs) could prove important in effective management of plant nutrients as an element of broader agricultural development. IPNSs are designed to balance the nutrients available to the farmer from all available sources for their optimal productive use, and to minimise "leakage" of minerals into the wider environment.

Systems are tailored to meet the needs of specific farming types, yield targets, the physical resource base, and the farmer's socioeconomic background. Implementing such systems would involve a formidable improvement in average levels of service to low-income farmers, in terms of technical advice, inputs, credit, marketing facilities, and public investment in agriculture.

10

Technological Advancements in Food Products

Carotenoids are responsible for the pleasing yellow, orange or red color of many foods. The role of some of these compounds as provitamin A precursors has been known for years. Other beneficial effects to human health have been more recently attributed to carotenoids, such as enhancement of the immune response and reduction of the risk of degenerative diseases such as cancer, cardiovascular diseases, cataract and macular degeneration, these actions not being restricted to the provitamins A.

Thus, the consumption of carotenoid-rich foods is widely recommended. The principal carotenoids of foods are β-carotene, β-cryptoxanthin, lycopene, lutein and violaxanthin. Except for violaxanthin, these are also the principal carotenoids found in the human plasma, and together with zeaxanthin, are the carotenoids most studied in terms of human health. β-Carotene and β-cryptoxanthin are provitamins A.

Lycopene is vitamin A inactive but is a more efficient antioxidant than β-carotene. It has been linked with reduction of the risk of cancer, especially lung, stomach and prostate cancer. Lutein and zeaxanthin are the carotenoids associated with lowered risk of macular degeneration and cataract.

Variation in the Carotenoid Composition

Foods vary qualitatively and quantitatively in their carotenoid composition. Dark green leafy vegetables, palm oil, palm fruits, carrot, orange sweet potato, mature squashes and pumpkins, tomato, yellow, orange and red tropical fruits are rich sources of carotenoids. In a given food, qualitative and especially quantitative differences also occur due to factors such as cultivar or variety, stage of maturity, climate or geographic site of production, part of the plant utilised, farming practice, processing and storage of food.

Differences among cultivars of the same food are well documented. An example is shown in Table 1. The red-fleshed papayas Solo, Formosa and Tailandia, produced in the Brazilian State of Bahia, differed particularly in the lycopene content with the Tailandia papayas having twice as much as the other two cultivars. The orange-fleshed common papaya is devoid of lycopene.

Table 1. Cultivar and geographic effects on the principal carotenoids (μg/g) of papaya.

Carotenoid	Common SP	Solo BA	Formosa SP	Formosa BA	Tailândia BA
β-carotene	1.2	2.5	1.4	6.1	2.3
β-cryptoxanthin	8.1	9.1	5.3	8.6	9.7
lycopene	-	21	19	26	40

SP – São Paulo, BA – Bahia

Maturation of a vegetable or ripening of a fruit is usually accompanied by enhanced carotenoid biosynthesis. This is shown in Table 2 for Tommy Atkins mangoes, in which the major carotenoids increased from the mature-green to the ripe. The same trend was observed in the cultivar Keitt.

Table 2. Principal carotenoids (μg/g) of Tommy Atkins mangoes (Mangifera indica) at three stages of maturity.

Carotenoid	Stage of maturity Carotenoid		
	Mature-green	Partially ripe	Ripe
β-carotene	2.0 ± 0.8	4.0 ± 0.8	5.8 ± 2.5
violaxanthin	6.9 ± 3.0	18 ± 7	22 ± 9
cis-violaxanthin	3.3 ± 1.3	9.0 ± 3.2	14 ± 5

In the same lot of acerola, ready for consumption, the firm ripe (red-orange) and fully ripe (darkred) fruits were separated and analysed separately. The composition differed especially in β- carotene, which was only 2.2 μg/g in the firm-ripe fruits and much higher (12 μg/g) in the fully ripe fruits (Porcu and Rodriguez-Amaya, unpublished results). Greater difference would be expected when comparison is made from the immature stage. This was indeed the case with Cucurbita moschata cultivar Menina Verde (Table 3). α-Carotene and β-carotene increased dramatically during maturation.

Table 3. Principal carotenoids (μg/g) of Cucurbita moschata Cv. Menina Verde at two stages ofmaturity.

Carotenoid	Stage of maturity Carotenoid	
	Immature	Mature
α-carotene	0.1	23
β-carotene	1.5	39

The β-carotene content of mature leaves of lettuce (12 μg/g) and endive (14 μg/g) were about three times those of the young leaves (3.5 and 4.2 μg/g, respectively). While fruits from cold regions are generally colored by anthocyanins, tropical and subtropical fruits owe their color mainly to carotenoids. Climatic effects are also seen in fruits of the same cultivars produced in regions of differing climates, elevated temperature and greater exposure to sunlight increasing carotenogenesis.

Comparing papayas of the cultivar Formosa produced in the hot Northeastern state of Bahia with those produced in the temperate Southeastern state of São Paulo, the former fruits presented higher β-carotene, β-cryptoxanthin and lycopene contents (Table 1). In Keitt mangoes, those produced in Bahia had more of the major carotenoids, especially β-carotene. In acerola, β-carotene and β-cryptoxanthin were much lower in the São Paulo fruits, compared to those from the neighboring hot Northeastern states of Pernambuco and Ceara which had similar composition.

Comparing kale at the same stage of maturity, produced in neighboring farms, all constituent carotenoids were significantly higher in samples collected from an organic farm than in those taken from a conventional farm using agrochemicals. This indicated that one of the chemicals used in the latter farm, probably herbicide, inhibited carotenoid biosynthesis in the leaves.

In many fruits, carotenoids are more concentrated in the peel than in the pulp. In Spondias lutea, for example, although β-carotene appeared slightly higher in the peeled fruit, β-cryptoxanthin was twice as much in the whole fruit (Table 4).

Table 4. Effect of peeling on the principal carotenoids (μg/g) of Spondias lutea and guava.

Fruit/Carotenoid	Whole fruit	Peeled fruit
Spondias lutea		
β-carotene	1.6 ± 0.2	2.6
β-cryptoxanthin	16 ± 2	8.3

Processing and Storage Effects

Many carotenogenic foods are seasonal and preservation at peak harvest is necessary to minimise losses, make the products available all year round and permit transportation to places other than the site of production. Processing and storage of foods should, however, be optimised to prevent carotenoid losses. Although industrial processing is more often focalised, losses on home preparation can also be, at times even more, considerable.

The tropical climate enhances carotenoid biosynthesis. On the other hand, this same ambient conditions may hasten destruction of carotenoids during post-harvest handling and storage. Carotenoid biosynthesis may continue, raising the carotenoid content, in fruits, fruit vegetables and root crops even after harvest, provided they are kept intact and not treated in any way that would inactivate the enzymes responsible for carotenogenesis. In leaves and other vegetables, post-harvest degradation of carotenoids appears to prevail, especially at high storage temperature and under conditions that favor wilting.

Carotenoids are naturally protected in plant tissues; cutting, shredding, chopping and pulping of fruits and vegetables increase exposure to oxygen and bring together carotenoids and enzymes that catalyse carotenoid oxidation. The stability of carotenoids differs in different foods, even when the same processing and storage conditions are used. Thus, optimum conditions for carotenoid retention during preparation/processing differ from one food to another. Carotenoids per se have different susceptibilities to degradation.

The major cause of carotenoid destruction during processing and storage of foods is enzymatic or non-enzymatic oxidation. Isomerisation of trans-carotenoids to the cis-isomers, particularly during heat treatment, alters their biological activity and discolors foods, but not to the same extent as oxidation. Enzymatic degradation of carotenoids may be a more serious problem than thermal decomposition in many foods.

In home preparation, losses of carotenoids generally increase in the following order: microwaving < steaming < boiling < sautéing. Deep-frying, prolonged cooking, combination of several preparation and cooking methods, baking and pickling all result in substantial losses of carotenoids. Whatever the processing method chosen, retention of carotenoid decreases with longer processing time, higher processing temperature and cutting or puréeing of the food. Reducing processing time and temperature, and the time lag between peeling, cutting or puréeing and processing improve retention significantly. High temperature, short-time processing is a good alternative.

The heat treatment in blanching may provoke some losses of carotenoids, but the inactivation of oxidative enzymes will prevent further and greater losses during holding before thermal processing, slow processing and storage. Freezing and frozen storage generally preserve the carotenoids, but slow thawing can be detrimental, particularly when the product has not been properly blanched. Peeling and juicing result in substantial losses of carotenoids, often surpassing those of heat treatment.

Traditional sun-drying, although the cheapest and most accessible means of food preservation in some regions, causes considerable carotenoid destruction. Drying in a solar dryer, even of simple and inexpensive design, can appreciably reduce losses. Protecting the food from direct sunlight also has a positive effect. Exclusion of oxygen (e.g. through vacuum or hot filling, oxygen-impermeable packaging, inert atmosphere), protection from light and low temperature diminish carotenoid decomposition during storage.

Alteration or loss of carotenoids during processing and storage of foods thus occur through physical removal (e.g. peeling), isomerisation and enzymatic or nonenzymatic oxidation, the last two reactions being due to the many double bonds in the carotenoid molecules. In programmes designed to promote production and consumption of carotenoid-rich foods, the following measures have been recommended:

— Identify locally available or potentially available food sources of carotenoids.

— Select and promote greater production of rich sources, which also have good agronomic, technological and sensory properties.

— Process carotenoid-rich food cultivars at optimum maturity/ripeness, stage at which they are both appropriate for processing and contain high levels of carotenoids.

— Avoid peeling foods. Losses during juicing should also be minimised.

— Consume or thermally process foods immediately after cutting, chopping or macerating. Consider blanching to inactivate the enzymes.

— Advocate simple measures such as avoiding prolonged cooking, cooking with the lid on, washing before peeling and cutting, avoiding cutting or macerating into very small pieces, keeping the food intact during storage, storing at low temperature in the dark and keeping storage time at a minimum.

— Optimise cooking, processing and storage conditions. Minimise cooking/processing time and temperature.

— Use solar-drying, in which the food is protected from direct sunlight, instead of traditional sun-drying.

For a long time the major concern about processing in relation to carotenoids had been preventing losses. In nature carotenoids are protected by the cellular structure, the destruction of which renders the carotenoids vulnerable to degradation. On the other hand, this natural protection lowers bioavailability.

Processing denatures proteins and breaks down the cell walls, making the release of carotenoids from the food matrix during digestion easier. It is thus recommended that strategies to increase the dietary intake of carotenoids include enhancement of bioavailability. Cooking/ processing conditions should be optimised so that appreciable losses of carotenoids are prevented while the bioavailability is increased.

11

Nutritional Quality

Nutritional Quality of Soyfoods

The market for soyfoods has grown exponentially as food processors and technologists have developed an increasing number of healthy, tasty and convenient foods. The growth in the sector is also due to the increasing number of consumers choosing to follow a vegetarian diet, but for others, it is to capitalise on the health benefits associated with soy and the bioactive components it includes. Interest in the physiological role of these bioactive compounds (phyto-chemicals) present in plants has also increased dramatically over the last decade.

Of particular interest is the role that the phytoestrogens can play in promoting health. Phytoestrogens are non-steroidal oestrogens, and contain the groups of compounds known as isoflavones, lignans and stilbenes. The richest food source of isoflavones is soybeans with smaller quantities found in other beans. Lignans are found in a range of cereals and vegetables, with flaxseed being a particularly concentrated source, and stilbenes—although widely found in nature—are usually only consumed in grapes, peanuts and their products.

Research and epidemiological data indicate that diets rich in phytoestrogens can exert a physiological effect in the body that promotes a healthy heart, blood vessels, bones and other tissues and can promote overall well being. There is no doubt that the most extensively researched plant protein containing bioactives is soy protein; each year hundreds of papers are added to the already immense scientific literature. It was this wealth of information that led the regulatory authorities in the States—the Food and Drug Administration (FDA)—to permit a health claim for soy protein.

In products containing a minimum of 6.25g soy protein/serving it can be claimed that "25g soy protein a day as part of diet low in fat and cholesterol may reduce the risk of heart disease". However, soy protein and it's naturally occurring isoflavones has health benefits much broader than heart health. Recent research has demonstrated a potential role in bone health and in the maintenance of healthy breasts and prostate.

Reducing the Risk of Chronic Diseases

Not one meta-analysis, but two have established the cholesterol-lowering attributes of soy protein, demonstrating the weight of scientific evidence that supports this claim. The average intake of soy protein was 47g (range 17 - 124g). 25 or 50g soy protein/day would reduce plasma cholesterol by 0.23 and 0.45mmol/l respectively. People with moderate high cholesterol (6.46mmol/l) should have even larger reductions. Since this time further studies monitoring the effects of soy protein have been published in the literature. Eight of the studies met the criteria adopted for the earlier meta-analysis and have been examined as a group.

More recent studies have tended to use lower quantities of soy protein and have adopted more similar experimental protocols. Hence, although the number of subjects is much smaller than the original meta-analysis (181 v 730), there is the benefit of less diverse, more consistent protocols. Confirmation that 25g soy protein can significantly reduce plasma LDL cholesterol by around 0.23mmol/l (6.1%) was revealed by researchers when they presented their recent meta-analysis of clinical studies at the 4th International Symposium on the Role of Soy in Preventing and Treating Chronic Disease, San Diego, USA.

The results from this analysis the pooled data and split by level of intake are shown below in table 1.

Table 1: The effect of soy protein on plasma cholesterol

Blood Parameter	Soyprotein g/day	Number of studies/subjects	Change mmol/l	Change %
LDL cholesterol	20 - 86	8/181	0.23	- 6.1
LDL cholesterol	25	3	0.26	- 6.2
LDL cholesterol	63	3	0.33	- 8.5

The data was also analysed by reference to its isoflavone content; low isoflavone (6mg) or high (46mg). The reductions in plasma LDL-cholesterol were 0.05 and 0.26 mmol/l respectively for the low and high group and the difference between groups was significant ($P>0.014$). The author concluded that 80% of the cholesterol lowering effect was related to isoflavone content.

Elevated blood cholesterol level is a major risk factor for heart disease, consequently much of the soy protein research has focused on this effect. However there is promising research indicating other heart health benefits. For example soy protein may be effective in:

— *Reducing LDL oxidation:* Isoflavones are antioxidants and it has been reported that they prevent lipoprotein oxidation that may minimise atherosclerosis (arterial disease), as oxidised LDL particles are considered more atherogenic.

— *Improving arterial condition:* There is a smaller body of evidence for a positive effect of soy phytoestrogens on atherosclerosis; they may have direct effects on the artery wall and the cells involved with the progression of the atherosclerotic plaque. For maximum benefit recent evidence has shown that soy should be incorporated into the diet in the early stages of lesion development. But the extent of improvement can be significant, animal data has shown a 70% lesser extent of coronary artery atherosclerosis when soy protein was included in the diet.

— *Thrombus development*: It has been shown that genistein interferes with platelet activation and aggregation, and inhibits thrombin action thereby reducing clot formation and the overall development of the thrombosis.

— *Reducing blood pressure:* Although the evidence is not extensive, recent information from clinical trials has shown that the inclusion of isoflavone rich soy can significantly reduce diastolic blood pressure in both healthy perimenopausal woman and middle-aged men and in those with elevated cholesterol.

— *Other cardiovascular factors:* for example a reduction in plasma homocysteine was recently reported when soy was included in the diet and direct oestrogen receptor (ERâ)-mediated effects on atherosclerosis inhibition.

Bone Health

Bone develops and matures as we get older, reaching its maximum, or peak bone mass at around the age of 30 years. Osteoporosis (the silent epidemic) affects 1 in 3 women and 1 in 12 men at some stage of their lives.

Oestrogen replacement therapy is the most effective way of retaining bone mass in older women, but compliance and take-up is poor. It is believed that the consumption of soy protein has a conservatory effect on calcium, this may be related to the protein content per se, or the isoflavone content of soy protein.

Certainly the amount and type of protein in the diet affects calcium excretion and bone resorption. Compared with animal protein, soy protein does not cause elevated calcium levels in urine and results in enhanced calcium retention. Recently, several clinical trials with women around the time of the menopause have been published where bone density or markers of bone turnover have been measured following the consumption of soy containing foods.

Despite considerable losses in bone mineral content (BMC) and bone mineral density (BMD) in the placebo group, those receiving dietary soy and isoflavones (soymilk) showed no change from zero. Indicating a reduced rate of bone loss in the soy group at this critical time. The effect on both BMC and BMD was higher among equol producers, for example, BMD was +0.6% and +1.1% in non-equol and equol producers respectively. Indicating that equol may play a positive role in bone turnover.

Although the evidence is not yet conclusive it appears that soy protein with isoflavones:

— Can be calcium sparing

— Can help moderate the bone loss after the menopause

— May be used to complement HRT

Components of Soy Protein

Protein is one of the essential nutrients in the diet. Both the amount and quality of protein provided by a food are important. There has been a long-standing (and incorrect) perception that soy protein is inferior to protein from animal sources. This misconception is largely due to the criteria traditionally used to evaluate protein quality.

The Protein Efficiency Ratio (PER) was the common method for evaluating protein quality. As knowledge about human amino acid needs increased, the PER's shortcomings were recognised, and in 1989 a new protein standard was introduced. Now all proteins around the world are evaluated according to this system.

In 1991 the Food & Agricultural Organisation (FAO) and World Health Organisation (WHO) recommended the Protein Digestibility-Corrected Amino Acid Score, or PDCAAS as a more accurate method for evaluating protein quality. PDCAAS takes several factors into account, including a food protein's essential amino acid content, its true digestibility, and its ability to supply essential amino acids in amounts adequate to meet human needs. In order to meet even the most stringent protein requirements, the PDCAAS compares the amino acid content of a food protein to the needs of a two-to-five-year old child, which represent the most demanding amino acid requirements of any group except infants.

The highest PDCAAS value that any protein can receive is 1.00. A score of 1.00 means that after digestion of the food protein, one unit of protein provides 100% of the indispensable amino acids required by a two-to-five-year old child. Scores higher than 1.00 are rounded down since any amino acids in excess of those required will not be utilised by the body as protein (that is, to build, repair, and maintain tissue) but, instead, will be deaminated and either used for fuel or stored. The PDCAA scores for soy protein and casein are the same, and are both 1.00.

Bioactive Glycoproteins

Soybean lectin is not heat stable unlike the lectins in other legumes, such as lima beans or kidney beans that are heat stable and growth inhibiting. Soybean lectins have a strong affinity for carbohydrates on the cell surface and this property results in a number of medical applications including fractionation of bone marrow cells, as a diagnostic agent in stomach cancer and in the treatment of ulcerative colitis. These properties related to commercial food preparations of soy that have only residual levels of lectins have not been investigated, although in view of their potential, merit further investigation.

Isoflavones

Interest in isoflavones increased, as it was believed that the health promoting properties of soy related to it's content of these compounds. And as analysis of blood and urine samples for isoflavones revealed that Japanese and Asian populations had an isoflavone content hundred-fold higher than Western populations.

Genistein, daidzein and glycitein are the specific isoflavone families found in soy and are present as 60%, 35% and 5% respectively of the total isoflavones. The ratio of the different families and proportion of the different forms within a family have an effect on the bioactivity of the isoflavones. The aglycone (without glucose) form of isoflavones is the most physiologically active.

The typical isoflavone content of a range of food products is given below in table 2. It can be seen that the isoflavone content of soy is variable depending on the variety, growing conditions, crop year and its subsequent processing.

Metabolism of Isoflavone

The first stage of metabolism of isoflavones requires

hydrolysis by the bacteria in the gut. In germfree animals, it has been demonstrated that there is no appearance of isoflavones in blood or bile.

Table 2. The isoflavone content of food products

Food	Total Isoflavones (μg/g)	Intake μg minimum	in average maximum	portion (g)
Soybeans	579 – 3812	34740	228720	(60)
Tofu	79 – 674	10270	87620	(130)
Soymilk	34 – 175	8500	43750	(250)
Soy sauce	13 – 75	65	375	(5)
Green split peas	73	2920	2920	(40)

Similarly the use of antibiotics results in a lack of their detection in the body or in urine. Absorbed isoflavones are transported to the liver by the hepatic portal vein where they are rapidly conjugated, predominately with glucuronic acid and to a lesser extent with sulphuric acid. In both blood and urine isoflavones are found primarily in the conjugated form. The half-life of circulating isoflavones is 7.9 hours. Consequently, to maximise the effective exposure to isoflavones then their consumption 2-3 times a day is desirable.

The isoflavone daidzein is converted to equol by bacteria in the gut. However, only around a third to a half of Western population are equol-producers. Equol when produced is biochemically active and circulates in the blood; about 50% of it circulates unbound and it has been shown to cross the placenta. It is excreted via the urine. In Western populations without soy in the diet then there is very little equol in the urine. Where it occurs, daidzein conversion to equol accounts for around 70% of its intake.

Three bacteria species responsible for the conversion of daidzein to equol were identified recently. However, whether there is also a family (phenotype) association with its production is currently under investigation. It is believed that it is beneficial to produce equol, as equol-producing women have steroid hormone profiles associated with lower risk of breast cancer, and have a greater lengthening of the follicular phase of the menstrual cycle.

In addition equol has a good affinity for ER, low affinity for lesion protein and is a better antioxidant then daidzein the isoflavone from which it is primarily derived. Recent evidence suggests that it may be this isoflavone which is responsible for the physiological properties of soy particularly those related to bone, breast and prostate health and menopausal symptoms.

The chemical structure of isoflavones and the mammalian oestrogen - oestradiol-17β share similar features - specifically the presence of a phenolic ring. In the body isoflavones can act either as oestrogens or oppose the action of oestrogens, which gives them a very special position. It has been suggested that soy-derived isoflavones may actually be naturally occurring selective oestrogen receptor modulators (SERMs).

Numerous other biological effects independent of ER have been established for isoflavones—antioxidant properties, reduced cell proliferation, and maintenance of blood vessel flexibility. The isoflavone threshold for biological effects has been identified as 30-50mg/day when incorporated in soy protein.

In many Western societies cancer is the second highest cause of premature death after disease and stroke. Interest in soy as a protective agent arose when it was identified that rates of and prostate cancer were much lower in Japan than in the West. Although there is mounting evidence to support that soy containing isoflavones has a protective effect on the development of some cancers, the number of reports are limited and there are inconsistencies in the data.

Epidemiological evidence suggests that the relatively low rates for colon, breast and prostate cancer in countries such as China and Japan, can at least, in part, be attributed to the regular lifetime intake of soyfoods. However, the epidemiology also indicates that adult consumption of soyfoods seems to confer little protection of postmenopausal breast cancer. It is difficult to carry out and interpret the findings of clinical studies in this area.

However, a recent case-controlled trial has looked at the urinary excretion of markers of breast cancer and indicated a reduced ratio of harmful compounds when soy isoflavones were introduced. This reduction was significant in the women in the group that were equol excretors. These observational findings suggest that equol metabolism is associated with a favourable alteration in oestrogen metabolism. Overall the evidence is far from conclusive although in those studies where there is a reduced risk, this is associated with a regular consumption of one serving of soy per day.

Although the role of soy in prostate cancer has been less extensively researched the epidemiology points to a protective role of soy isoflavones, but there is a lack of clinical data to support this contention.

12

Nutritional Quality and Modern Biotechnology

After water, food is our most vital item of consumption. About three-quarters of the human food supply comes from agriculture. Agriculture's traditional role of providing food, feed, and fiber is being augmented by biotechnology. Biotechnology will be a critical element in the development of crops, foods, and ingredients with traits with improved nutritional properties.

Developing plants with these improved traits involves overcoming a variety of technical challenges inherent in metabolic engineering programmes. Both traditional plant breeding and biotechnology-based techniques are needed to produce plants with the desired quality traits. Continuing improvements in molecular and genomic technologies are contributing to the acceleration of product development.

Plasticity of Plant Metabolism

Plants are remarkable in their ability to synthesise a variety of organic compounds, such as vitamins, sugars, starches, fatty acids, and amino acids. As many as 80000 to 100000 different substances are synthesised in plants, including macronutrients, micronutrients, antinutrients, allergens, endogenous toxicants, and other plant-specific compounds that are significant to human and animal health.

This plasticity is elegantly demonstrated in the way that plants respond to environmental stimuli such as pathogen attack. Functional complexity begins with the exogenous signals perceived from the pathogen, continues with the mechanisms of signal perception and signal transduction, and results in extensive "reprogramming" of cellular metabolism, involving extensive changes in gene activity. Thus, pathogen defense entails a major shift in metabolic activity, rather than altered expression of a few unique, defense-related genes.

The observed complexity serves as a paradigm of the flexibility and plasticity of plant metabolism. Many of these same metabolites have either positive or negative impacts on the nutritional characteristics of plants. For example, the shikimate pathway includes a number of phytochemicals that can have either good or bad effects. These compounds include phenylpropanoids, coumarins, stilbenes (some such as resveratrol are beneficial, while others such as kawain have negative effects), flavonoids, and tannins.

The next generation of plants will focus on value-added output traits where valuable genes and metabolites will be identified and isolated, with some of the metabolites being produced in mass quantities for niche markets. The nutritionally improved crops in the current development pipeline will be well understood and well characterised from a compositional perspective as they undergo safety and nutritional assessment following existing regulations that are more than adequate to address any potential concerns.

However, some of the more potentially beneficial modifications will require more thorough understanding of plant metabolism and methods to achieve effective changes in the desired metabolic endpoints. Although progress in dissecting metabolic pathways and our ability to modify gene expression in GM plants has been most impressive during the past 2 decades, attempts to use these tools to engineer plant metabolism have met with more limited success.

Metabolic engineering typically involves the redirection of cellular activities by the modification of the enzymatic, transport, and regulatory functions of the cell using recombinant DNA (rDNA) and other techniques. Since the success of this approach hinges on the ability to change host metabolism, its continued development will depend critically on a far more sophisticated knowledge of plant metabolism, especially the nuances of interconnected cellular networks, than currently exists.

Although the enzymological sequences and intermediates of many metabolic pathways in small number of well-studied organisms are known with some confidence, little is known in quantitative terms about the controls and integration of these pathways. The necessary knowledge also includes conceptual and technical approaches necessary to understand the integration and control of genetic, catalytic, and transport processes. Though there are notable exceptions, most successful attempts at metabolic engineering thus far have focused on modifying (positively or negatively) the expression of single genes (or a series of individual enzymatic steps) affecting pathways.

Generally, more success has been achieved when conversion or modification of an existing compound to another has been targeted than when an attempt has been made to significantly change flux through a pathway (for example, increasing the oleic acid concentration in canola oil, as will be discussed later). Attempts to modify storage proteins or secondary metabolic pathways have also been more successful than have alterations of primary and intermediary metabolism.

Research to improve the nutritional quality of plants has historically been limited by a lack of basic knowledge of plant metabolism and the stimulating challenge of resolving complex interactions of thousands of metabolic pathways. With the tools now being harnessed through the fields of genomics and bioinformatics, there is the potential to identify genes of value across species, phyla, and kingdoms. Through advances in proteomics, it is becoming possible to quantify simultaneously the levels of many individual proteins and to follow posttranslational alterations that occur in pathways. Metabolomics allows the study of both primary and secondary metabolic pathways in an integrated fashion. With these evolving tools, a better understanding of global effects of metabolic engineering on metabolites, enzyme activities, and fluxes is beginning to be developed.

The increase in our basic knowledge of plant metabolism during the coming decades will provide the tools necessary to modify more effectively the nutritional content of crops to have a positive effect on many aspects of human and animal health. In addition to metabolic considerations, attention needs to be given to the site of synthesis and site of activity of an enzyme. Signal sequences or transit peptides coding sequences attached to introduced genes are not always sufficient to ensure appropriate targeting. For example, charge and size of a protein may affect the efficiency of transportation into plastids.

Another complexity found in biological systems is redundancy of pathways and the ability of plants to compensate as they often contain more than one enzyme capable of catalysing a similar reaction. A potential approach to counter some of these problems in metabolic engineering of pathways involves the manipulation of transcription factors that control networks of metabolism. For example, expression of maize transcription factors C1 and R, which regulate production of flavonoids in maize aleurone layers, together under the control of a strong promoter resulted in a high accumulation rate of anthocyanins in Arabidopsis, presumably by activating the entire pathway.

Such expression experiments hold promise as an effective tool for the determination of transcriptional regulatory networks for important biochemical pathways. In summary, metabolic engineers must not only understand the fundamental physiology of the process to be impacted, but also the level, timing, subcellular location, and tissue or organ specificity that will be required from a transgene to ensure successful manipulation of that physiology.

Gene expression can be modulated by numerous transcriptional and posttranscriptional processes. Correctly choreographing these many variables is the element that makes metabolic engineering in plants so challenging. In conjunction with such increases in the understanding of plant metabolism, the challenge then remains to understand how components in the diet interact with human or animal metabolism to benefit their health and well-being. This challenge is at least as complex as the task of increasing or decreasing the amount of a specific protein, fatty acid, or other component of the plant itself. It is of little use producing a plant with a supposed nutritional benefit unless that benefit actually improves the health of humans or animals.

Metabolic engineering is generally defined as the redirection of one or more enzymatic reactions to improve the production of existing compounds, produce new compounds, or mediate the degradation of compounds. Substrate-product relationships in plant pathways were initially elucidated through the application of radiolabel tracer studies during the 1960s and 1970s.

In the 1980s, with the advent of rDNA technology, tools such as cloning, promoter analysis, protein targeting, plant transformation, and biochemical genetics were developed. The GM crops with improved agronomic traits presently being grown on more than 60 million ha around the world are a product of the application of these technologies to crop plants. These products provide benefits to the farmer and community in reducing insecticide and herbicide usage and increasing the ability of farmers to conserve soil and other resources. They generally involve the relatively simple task of adding a single gene or small number of genes to plants. These genes in the main function outside of the plant's primary metabolic processes and thus have little or no effect on the composition of the plants.

Metabolic pathway modifications are complex, and the state of understanding of plant metabolism is sometimes insufficient to bridge the gap between the ability to clone, study, and modify individual genes and proteins and the understanding of how they are integrated into and affect the complex metabolic networks in plants. Regulatory oversight of such products has been designed to detect such unexpected outcomes and to ensure that products from GM plants are safe before they are commercialised.

Genomics-based strategies for gene discovery, coupled with high-throughput transformation processes and miniaturised automated analytical and functionality assays, have accelerated the identification of product candidates. Identifying rate-limiting steps in synthesis could provide targets for genetically engineering biochemical pathways to produce augmented amounts of compounds and new compounds. Targeted expression will be used to channel metabolic flow into new pathways, while gene-silencing tools can reduce or eliminate undesirable compounds or traits, or switch off genes to increase desirable products.

In addition, molecular marker-based breeding strategies have already been used to accelerate the process of introgressing trait genes into high-yielding germplasm for commercialisation. Analysis of fluxes in metabolic pathways in response to an environmental or genetic manipulation can help identify rate-limiting steps. Traditional biochemical hallmarks of potential regulatory, or rate-controlling, enzymes are that they catalyse reactions and are regulated by appropriate effector molecules.

The modification of enzymes of the carbon cycle to study their role in regulating pathway flux has provided some of the more interesting results from metabolic engineering studies in plants. For example, when the highly regulated Calvin cycle enzymes, fructose-1, 6-bisphosphatase and phosphoribulokinase, were reduced 3- and 10-fold in activity, respectively, minor effects on the photosynthetic rate were observed.

In contrast, a minor degree of inhibition of plastid aldolase, which catalyses a reversible reaction and is not subject to allosteric regulation, led to significant decreases in photosynthetic rate and carbon partitioning. Thus aldolase, an enzyme seemingly irrelevant in regulating pathway flux, was shown to have a major influence over the pathway. Understanding of the individual kinetic properties of such key enzymes may not always be sufficient to understand their wider role in central metabolism.

Functional Foods

Functional foods are intended to appeal to consumers by

offering potential health benefits that go beyond satisfying basic nutritional needs. These foods exploit the growing scientific evidence supporting the role of a diet containing certain types of foods or phytochemicals in the prevention and treatment of disease. Epidemiological research has shown a positive association between dietary intake of food components found in fruits, vegetables, grains, fish oil, and legumes and their effect on chronic disease.

Beyond understanding of plant metabolism, the challenge then remains to better understand how components in the diet interact with human or animal metabolism to benefit their health and wellbeing. Although there exists extensive research and clinical support for specific nutrient effects as documented in the following sections, improving our knowledge at the fundamental level of molecular effects will better inform the decisions being made with respect to nutritional quality improvement.

This challenge is at least as complex as the task of increasing or decreasing the amount of a specific protein, fatty acid, or other component of the plant itself. It is of little use producing a plant with a supposed nutritional benefit unless that benefit can be translated into positive health or nutritional impacts in humans or animals. The application of rDNA technology to improve plant-specific components known to have benefit for human health that goes beyond meeting basic nutritional requirements is one way to introduce new functional foods into the marketplace.

In addition to functional foods, rDNA technology allows the engineering of plants to address issues of animal nutrition and the impact of animal effluent on the environment. A good example of this is the addition of phytase enzymes to crops to reduce the need to add phosphorus to feed. Most of the phosphorus is added because the phosphorus in phytic acid is not bioavailable and because of the sequestering effect of phytic acid on uptake of divalent mineral ions.

Modifications

Proteins and Amino Acids

Humans, as well as poultry, swine, and other nonruminant animals, have specific dietary requirements for each of the essential amino acids. A deficiency of 1 essential amino acid limits growth and can be fatal. In animal feeds, the primary limitations of maize and soybean meal-based diets are for lysine in nonruminant mammals and methionine in avian species. Maize with increased levels of lysine and soybeans with increased levels of methionine could allow diet formulations with improved amino acid balance, without the need to add crystalline lysine and methionine.

Most plants have a poor balance of essential amino acids relative to the needs of animals and humans. The cereals tend to be low in lysine, whereas legumes (soybean, peas, and so on) are often low in the sulfur-rich amino acids methionine and cysteine. Successful technical examples to date to enhance free amino acids levels include high-lysine maize and high-lysine canola and soybeans.

In areas such as less-developed countries, where it is difficult to obtain access to the components necessary for a balanced diet, these types of modifications could offer a particular advantage. Consumption of foods prepared from these crops potentially can help prevent protein malnutrition in such regions, especially among children, as well as increase the availability of animal protein in developing countries by improving the quality of animal feed. From an engineering perspective, one of the most straightforward methods to modify amino acid compositions of food and feed is by expressing proteins with high levels of the desired amino acids in the seed (the major food source).

One method of modifying storage protein composition is to introduce heterologous or homologous genes that code for proteins containing elevated levels of sulfur-containing amino acids (methionine, cysteine) and lysine. These proteins can be from other natural sources or can be synthetic. Transgenic plants exhibited a 2- and 5-fold increase in the total protein content in leaves and roots, respectively, over that of control plants.

A significant increase in the level of essential amino acids such as methionine, threonine, tryptophan, isoleucine, and lysine was also observed. An example of the use of proteins from natural sources is the work of Chakraborty and others, who reported introducing an albumin gene for a nonallergenic protein from Amaranthus, rich in all essential amino acids, into potato.

The resulting tuber composition corresponds well with the World Health Organisation (WHO) standards for a nutritionally rich protein for optimal human nutrition. In this case, there was a striking increase in the growth rate and production of tubers in transgenic populations compared to the control. There was also an increase in the total protein content, with an increase in most essential amino acids. The results of these experiments document, in addition to successful nutritional improvement of potato tubers, the feasibility of genetically modifying other non-seed food crop plants with novel protein composition.

An important issue is that of ensuring that the total composition of storage proteins, for example, is not altered to the detriment of the development of the crop plant when attempting to improve amino acid ratios. Modifying soybean storage pro-teins in such a way that the 3-dimensional structure is maintained, and so that the modified proteins can accumulate in the seed at levels comparable to the endogenous seed proteins. Maize produces a methionine-rich protein (delta-zein) in the grain but at a low level.

Dzr1,a protein, that binds an intronic region and degrades delta-zein mRNA before translation. They replaced the targeted intronic region with an intron from another maize gene. This prevented Dzr1 from degrading delta-zein RNA and maximised the production of the methionine-rich protein. Chickens fed diets containing this maize grew significantly faster than chickens fed conventional maize. This modification could potentially save animal farmers $1 billion per year in synthetic methionine supplements to maize-based feed.

Attempts to manipulate the free lysine content of seeds illustrate that one needs to consider catabolic, as well as anabolic, variables when trying to engineer a particular metabolic phenotype in plants. A key step in lysine synthesis is catalysed by dihydrodipicolinate synthase (DHDPS), which is feedback inhibited by the pathway endproduct (lysine) and, thus, plays a key role in regulating flux through the pathway. Engineering plants to overexpress a feedback-insensitive bacterial DHDPS greatly increased flux through the lysine biosynthetic pathway.

However, in most cases this did not result in greater steady-state lysine levels because the plants also responded by increasing flux through the lysine catabolic pathway through elevation of lysine-ketoglutarate reductase. Substantial increases in lysine only occurred in plants where flux increased to such a level that the first ensyme of the catabolic pathway became saturated, again illustrating the potential complexities of metabolic regulation.

Carbohydrates

Plants make both polymeric carbohydrates and individual sugars. The biosynthesis of these compounds is sufficiently understood to allow the bioengineering of their properties and to engineer crops to produce polysaccharides not normally present. The term prebiotic is used to describe an indigestible food ingredient, such as fructooligosaccharides (FOS), that beneficially affects the microflora by selectively stimulating the growth and/or activity of beneficial bacteria.

Fructans (plant inulins) and fructooligosaccharides may be important ingredients in functional foods, because evidence suggests that they promote a healthy colon and help reduce the incidence of colon cancer. The FOS may have anticarcinogenic, antimicrobial, hypolipidemic, and hypoglycemic actions in some. They may also help improve mineral absorption and balance, and may have antiosteoporotic and antiosteopenic activities.

Inulins are only slightly digested in the small intestine. They are, however, fermented by a limited number of colonic bacteria. This could lead to changes in the colonic ecosystem in favor of some bacteria, such as Bifidobacteria, which may have health benefits. Oral administration to humans of fructans, such as oligofructose and inulin, has been shown to increase the number of bifidobacteria in stools.

Bifidobacteria may inhibit the growth of pathogenic bacteria, such as Clostridium perfringens and diarrheogenic strains of Escherichia coli. Inulins are considered to be bifidogenic factors. Their energy content is about half that of digestible carbohydrates or about 1 to 2 kcal/g. The possible anticarcinogenic activity might be accounted for, in part, by the possible anticarcinogenic action of butyrate. Butyrate, along with other shortchain fatty acids, is produced by bacterial fermentation of FOS in the colon.

Some studies have shown that butyrate induces growth arrest and cell differentiation and may also upregulate apoptosis, 3 activities that could be significant for antitumor activity. The FOS may lower serum triglyceride levels in some individuals. The mechanism of this possible effect is unclear. Decreased hepatocyte triglyceride synthesis is a hypothetical possibility. The FOS may also lower total cholesterol and LDL-cholesterol levels in some people. Again, the mechanism of this possible effect is unclear. Propionate, a product of FOS fermentation in the colon, may inhibit HMG-CoA reductase, the rate-limiting step in cholesterol synthesis. Thus, there is interest in modifying plants to produce this polymeric carbohydrate.

The main crop of interest for producing fructan is the sugar beet because the major storage component of this species is sucrose, the direct precursor for fructan biosynthesis. The soluble oligosaccharides, stachyose and raffinose, which are found in soybeans, are not digested and can cause flatulence and digestive problems, producing discomfort in humans. These compounds in conventional soybean or soybean meal are similarly not digested by nonruminant animals, resulting in reduced feed efficiency.

Researchers found that the incorporation of lowstachyose soybean meal from nonmodified sources in prestarter pig diets tended to improve growth performance. In addition, the increased sucrose content of lowstachyose soybean results in foods with a sweeter taste than do their traditional counterparts. Manipulating the level of this family of oligosaccharides through rDNA technology has been achieved by inhibiting galactinol synthase activity. This is the first committed step in the pathway and involves the synthesis of galactinol from UDP-Gal and myo-inositol. The individual members are then synthesised by distinct galactosyl transferases.

As raffinose and stachyose may be crucial during seed development and storage, which is based on the transfer of α-galactosidase from a thermophilic bacterium (Thermotoga neapolitana) into legumes and inducing α-galactosidase to degrade the oligosaccharides after harvesting by changing the temperature. Starch is an important storage polysaccharide in many plants. It is composed of densely packed α-glucans, consisting of α-1,4-and α-1,6-linked glucose residues. Engineering starch content and composition in potatoes is of interest. Plant ADP glucose pyrophosphorylase (ADPGPP) is sensitive to allosteric effectors and has been proposed to be a key regulator of starch biosynthesis. The higher starch content results in far less fat absorption during frying, because the moisture lost during frying is replaced by oil. However, there are still problems of irregular granule distribution throughout the tuber to be solved.

Fiber and Lignans

Fiber is a group of substances chemically similar to carbohydrates, except that nonruminant animals poorly digest fiber. Fiber provides bulk in the diet, such that foods rich in fiber are satisfying without contributing significant calories. Current controversies aside, there is ample scientific evidence to show that prolonged intake of foods high in dietary fiber has various positive health benefits in humans, especially the potential for reduced risk of cardiovascular disease and colon and other types of cancer.

The relative risk for fatal myocardial infarction was 0.45 among men with the highest intake of fiber (median 28.9 g/d) compared with men with lowest intake of fiber. Fiber type and quantity are undoubtedly under genetic control, although this topic has received little attention. The technology to modify fiber content and type by genetic engineering would be a great benefit in persuading the many individuals who, for taste or other reasons, do not include adequate amounts of fiber in their daily diet. For example, fiber content could be added to more preferred foods or the more common sources of dietary fiber could be altered for greater health benefits.

Other fiber-associated compounds include lignans. The 2 lignans of primary interest in mammals, enterodiol and its oxidation product, enterolactone, are formed in the intestinal tract by bacterial action on plant lignan precursors. Flaxseed is the richest source of mammalian lignan precursors. Because enterodiol and enterolactone are structurally similar to both naturally occurring and synthetic estrogens, and have been shown to possess weakly estrogenic and antiestrogenic activities, they may play a role in the prevention of estrogen-dependent cancers.

Genes encoding all the enzymes for the conversion of coniferyl alcohol (lignan and lignin precursor) to secoisolariciresinol, a major dietary phytoestrogen, have been cloned. Other alcohol derivatives such as plant sterols (mainly sitostanol) exhibit a dose-dependent action inhibiting cholesterol absorption while increasing cholesterol excretion and upregulating cholesterolgenesis in hamsters, resulting in lower circulating lipid levels. However low-fiber feedstuffs are often favored for nonruminant animals. Nonruminant animals do not produce enzymes necessary to digest cellulose-based plant fiber.

Plants low in fiber should yield more digestible and metabolisable energy and protein and less manure and methane when fed to these species. US Dairy Forage Center (USDFRC) estimates that a 10% increase in fiber digestibility would result in an annual $350 million increase in milk/beef production and decreased generation of manure, manure, USDFRC estimates that a 10% increase in fiber digestibility is equivalent to 2.8 million tons decrease in manure solids produced each year.

Improved digestibility of livestock feed is therefore highly desirable. The altered lignin content and composition resulted in increased rate and extent of rumen digestion. Researchers identified and cloned a brown midrib (Bmr) gene, which encodes caffeic acid O-methyltransferase (COMT), a lignin-producing enzyme. They generated mutants that give rise to plants that contain significantly lower lignin in their leaves and stems, leading to softer cell walls compared to wild type.

The plant-softening mutations improve the digestibility of the food, and livestock seem to prefer the taste. Such improved fiber digestibility in nonruminants should have significant beneficial effects because the efficiency of digestion of most highfiber diets for nonruminants is far from optimised.

Oils and Lipids

Gene technology and plant breeding are combining to provide powerful means for modifying the composition of oilseeds to improve their nutritional value and provide the functional properties required for various food oil applications. The technology also has the potential to produce industrial oils and chemicals in genetically engineered crops. Genetic modification of oilseed crops can provide an abundant, relatively inexpensive source of dietary fatty acids with wide-ranging health benefits.

Production of lipids shown to have health benefits in vegetable oil provides a convenient mechanism to deliver healthier products to consumers without the requirement for significant dietary changes. The lipid biosynthetic pathway was one of the earliest pathways to be targeted for modification, and it represents one of the better examples of metabolic engineering in plants to date. Most enzymes required for fatty acid synthesis in plants have been cloned, and various academic and industrial groups have modified their expression to manipulate oilseed fatty acid composition.

Major alterations in the proportions of individual fatty acids have been achieved in a range of oilseeds using conventional selection, induced mutation, and, more recently, posttranscriptional gene silencing. Examples of such modified oils include low- and zero-saturated fat soybean and canola oils, canola oil containing medium chain fatty acids (MCFA), high-stearic acid canola oil (for trans fatty acid-free products), high-oleic acid (monounsaturated) soybean oil, and canola oil containing the polyunsaturated fatty acids (PUFA), γ-linolenic, stearidonic acids, and other omega-3 fatty acids. Altering the chain length and saturation level of the fatty acids can improve the nutritional qualities of some oils.

In addition, genes from various plant species may be introduced to produce unusual fatty acids in oilseed crops. LauricalTM, canola oil with high amounts of lauric acid, was the first commercial GM food oil. In this case, lauroyl-ACP thioesterase genes from the California bay laurel were cloned and transferred to canola (low-erucic acid rapeseed) oil crops.

In 1995, the FDA completed its evaluation of Laurical for use in food products. Medium chain fatty acids (MCFA) range from 6 to 10 carbons long and are only minor components of natural foods. The medium chain triglycerides (MCT) with these MCFA aid in absorption of calcium and magnesium and are rapidly oxidised as a quick source of energy. When MCT are substituted for long chain triglycerides (LCT) in the diet, animals gain less weight, store less adipose tissue, and experience an increase in metabolic rate.

Mice fed diets with MCT have also been shown to possess increased endurance in swimming tests over that of mice fed diets with LCT. Medium chain triglyceride oil has been included in medical foods, ergogenic aids, and dietary supplements. Because MCT are not readily available in high quantities in ordinary foods, they must be produced synthetically, making them of great interest to researchers.

Thus, researchers have used the morilena mushroom and plants to identify enzymes involved in production of the MCT capric and caprylic acid. Expression of an acyl-ACP thioesterase cDNA from C. hookeriana in seeds of canola, an oilseed crop that normally does not accumulate any capric and caprylic acid, resulted in a large increase in the levels of these 2 MCT. This illustrates the capacity to harness, through biotechnology, the genes contributing to phytochemical biodiversity in wild species and offers significant potential in the treatment of disease where such phytochemicals have proven health benefits.

Many types of fats are important, and the following sections will discuss different types of modifications with differing health implications. Edible oils rich in monounsaturated fatty acids provide improved oil stability, flavor, and nutrition for human and animal consumption. Oleic acid, a monounsaturate, can provide more stability than the polyunsaturates, linoleic and linolenic acids.

Higher monounsaturates are also preferred from a health perspective. Antisense inhibition of oleate desaturase expression in soybean resulted in oil that contained >80% oleic acid and had a significant decrease in polyunsaturated fatty acids. Clemente achieved a more stable effect using termination of transcripts with a self-cleaving ribozyme to enhance nuclear retention and serve as a tool to decrease specific plant gene expression achieving greater than 85% oleic, and saturated fatty acids levels of less than 6%.

High-oleic soybean oil is naturally more resistant to degradation by heat and oxidation, and so requires little or no postrefining processing (hydrogenation), depending on the intended vegetable oil application. While many lipids have important health implications, the longchain polyunsaturated fatty acids (PUFA), especially the omega-3 fatty acids found in fish, eicosapentaenoic acid (EPA) and docosahexaenoic acid (DHA), which are present in the retina of the eye and cerebral cortex of the brain, are some of the most well documented from a clinical perspective.

Docosahexaenoic acid is also the predominant structural fatty acid in the gray matter of the brain. It is believed that EPA and DHA play an important role in the regulation of inflammatory immune reactions and blood pressure, treatment of conditions such as cardiovascular disease and cystic fibrosis, brain development in utero, and, in early postnatal life, the development of cognitive function. Omega-3 fatty acids also seem to be beneficial in certain neuropsychiatric illnesses such as bipolar disorder, schizophrenia, and depression.

Current Western diets tend to be relatively high in omega-6 fatty acids and relatively low in omega-3 fatty acids. This is due in part to our high intake of vegetable oils that are rich in omega-6 fatty acids, and our low intake of oils and foods rich in omega-3 fatty acids, such as canola oil, flaxseed oil, or fatty fish. In plants, the microsomal ω-6 desaturase-catalysed pathway is the primary route of production of polyunsaturated lipids.

Vitamins and Minerals

For selected minerals and a limited number of vitamins, the clinical and epidemiological evidence is clear that they play a significant role in maintenance of optimal health and are limiting in diets worldwide. In addition, there is a growing knowledge base indicating that elevated intake of specific vitamins and minerals may reduce the risk of diseases such as certain cancers, cardiovascular diseases, and chronic degenerative diseases associated with aging.

Because of the difficulty in separating individual nutrient effects from an overall dietary pattern that may be fundamental to achieving these health benefits, improved dietary patterns should still be encouraged. If nutrient intakes associated with optimal health benefits are not achievable by dietary modification alone, fortification of foods will be an alternative route. Genetic engineering is a potentially important route of fortification, particularly since it would seem to avoid many of the technical problems associated with food fortification such as uneven distribution of minute quantities of nutrients, unstable mixing and settling, over- or underaddition, and so on.

Various groups (for example, the Consultative Group on International Agricultural Research) are using both traditional breeding and recombinant DNA approaches to develop biofortified crops that will be especially valuable in developing countries. Rice is a staple that feeds nearly half the world's population, but milled rice does not contain β-carotene or significant amounts of its precursors. Integrating observations from prokaryotic systems into their work has enabled researchers to clone the majority of the carotenoid biosynthetic enzymes from plants during the 1990s.

Research team at ETH-Zurich reported that immature rice endosperm is capable of synthesising the early intermediate of β-carotene biosynthesis. Using carotenoid pathway genes from daffodil and Erwinia and a Rubisco transit peptide, his team succeeded in producing β-carotene in the rice endosperm. This major breakthrough in the modified rice plant led to the development of "Golden indica Rice" based on the concept reported earlier, which showed that an important step in provitamin A synthesis can be engineered into a non-green plant part that normally does not contain carotenoid pigments.

Iron is the most commonly deficient micronutrient in the human diet, and iron deficiency affects an estimated 1 to 2 billion people. Anemia, characterised by low hemoglobin, is the most widely recognised symptom of iron deficiency, but there are other serious problems such as impaired learning ability in children, in-creased susceptibility to infection, and reduced work capacity.

Nutraceuticals

The search for new compounds to treat human disease has led to the formation of specialised biotechnology firms searching for nutraceuticals. The recommended dietary allowances do not reflect the growing knowledge base, which indicates that elevated intakes of specific vitamins and minerals significantly reduce the risk of diseases such as certain cancers, cardiovascular diseases, and chronic degenerative diseases associated with aging.

To obtain such therapeutic levels in the diet, additional fortification of the food supply will be required as well as modification of dietary preferences, or direct modification of micronutrient levels in food crops. Studies demonstrate that maximised dietary intake is not always correlated with optimised dietary benefit. Quercetin is a flavonoid that has been demonstrated in some studies to work optimally at very low concentrations in protecting against cancerous cell proliferation and the actions of the carcinogen PhIP (2-amino-1-methyl-6-phenylimidazo[4,5-b]pyridine) found in cooked meat.

After activation in the liver, PhIP can attack DNA to form DNA adducts. Using accelerator mass spectrometry (AMS), this group has shown that both quercetin and sulforaphane can inhibit DNA adduct formation in a dose-dependent manner. The protective mechanism of quercetin is through the inhibition of the phase I enzyme CYP 1A2, while sulforaphane acts through the induction of phase II detoxification enzymes such as glutathione transferases and UDP-glucuronosyl transferases. They further found that quercetin could ameliorate the effects of PhIP optimally at very low concentrations. As the concentration was increased, the effect was attenuated.

Similar effects may be found for other phytochemicals. This also illustrates the importance of taking a cautious approach to any research to increase phytochemicals with putative beneficial effects under the premise of "more is better." Unlike vitamins and minerals where mode of action is known, the primary evidence for the health-promoting roles of phytochemicals comes from epidemiological studies, and the exact chemical identity of many active compounds has yet to be determined.

However, for select groups of phytochemicals, such as non-provitamin A carotenoids, glucosinolates, and phytoestrogens, the active compound or compounds have been identified and rigorously studied. Other targets include improved iron content, through the production of iron-rich storage protein, bioavailable phosphorus released from phytate, and isoflavonoids. Other interesting products in the carotenoid pathway include lycopene, which may benefit the cardiovascular system by reducing the amount of oxidised low-density lipoprotein (LDL).

Recent epidemiologic studies have suggested a potential benefit of this carotenoid in reducing the risk of prostate cancer, particularly the more lethal forms of this cancer. Five studies support a 30% to 40% reduction in risk associated with high tomato or lycopene consumption in the processed form in conjunction with lipid consumption, although other studies with raw tomatoes were not conclusive.

Stilbenes, including resveratrol (3,5,4'-trihydroxystilbene), are phenolic natural products that accumulate in a wide range of plant species, including pine, grapevine, peanut, and rhubarb milled rice. Grapes and related foods, such as raisins and red wine, are among the few human dietary sources of resveratrol. This compound has attracted considerable notice as a substance with possible beneficial effects on human health.

An excellent antioxidant, resveratrol inhibits platelet aggregation and eicosanoid synthesis and is thought to contribute to improved heart function and lower blood cholesterol, based on epidemiological studies. It was shown to have "chemo-preventive" activity, preventing the formation of tumors in mouse skin bioassays, and, therefore, may help reduce cancer rates in humans.

Other phytochemicals of interest include flavonoids, such as tomatoes expressing chalcone isomerase that show increased contents of the flavanols rutin and a kaempferol glycoside; glucosinolates and their related products such as indole-3 carbinol (I3C); catechin and catechol; isoflavones, such as genistein and daidzein; anthocyanins; and some phytoalexins.

Antinutrients

Reducing phytate is an example of a biotechnology approach that solves both a nutritional and an environmental problem. Seeds store the phosphorus needed for germination in the form of phytate, a sugar alcohol molecule having 6 phosphate groups (inositol hexaphosphate). However, phytate is an antinutrient because it strongly chelates iron, calcium, zinc, and other divalent mineral ions, making them unavailable for digestive uptake.

Nonruminant animals generally lack the phytase enzyme needed for digestion of phytate. Poultry and swine producers in most countries currently add mined and processed (powdered) phosphate to the diets of their animals to enable optimal growth. Excess phosphate is excreted into the environment, resulting in water pollution. When low-phytate soybean meal is utilised along with low-phytate maize for animal feeds, the phosphate excretion in swine and poultry manure is reduced by half.

A series of GM rice lines have been developed to solve this problem. In addition, low-phytate maize was commercialised in the USA in 1999. Research indicates that the protein in low-phytate soybeans is also slightly more digestible than the protein in traditional soybeans. A number of studies have shown that optimal performance and bone mineralisation can result from diets without added phosphorus when phytase is included.

Researchers demonstrated that phytase supplementation to low-phosphorus diets improved performance, mineral use, tibia weight, and relative liver weight in broiler chickens fed different levels of phosphorus. Phytase supplementation of low-phosphorus diets improves performance, phosphorus digestibility, and bone mineralisation and reduces phosphorus excretion in pigs. Poultry grew well on the engineered alfalfa diet without any inorganic phosphorus supplement.

Thus, phosphorus supplements may be eliminated from poultry feed to reduce costs and reduce pollution. Other antinutrients that are being examined as possible targets for reduction are trypsin inhibitors, lectins, and several other heatstable components found in soybeans. Consideration must be given to possible increased susceptibility to pests and diseases when natural toxicants are removed, so the base germplasm should have input traits to counter this.

Reducing the amounts of trypsin inhibitors in soybeans would have a positive effect on the domestic feed industry and offer a competitive advantage for onfarm feeding of this protein source. If this can be combined with increases in the amounts of essential amino acids, very large improvements in productivity may be achieved.

Causes of Food Intolerance

While symptoms of food intolerance are common, true food allergy is less common. A food allergy is distinguished from food intolerance and other disorders by the production of antibodies (IgE) and the release of histamine and similar substances. The best-characterised true allergens include the superfamily cupins, which include globulins found in nuts and beans and albumins in nuts, and the superfamily prolamins found in cereal grains.

Other common allergens are hevein (initially from rubber trees), which causes contact dermatitis from latex, and chitinases. Foods that frequently cause malabsorption or other food intolerance syndromes other than direct IgE immune responses include wheat and other gluten-containing grains and cow's milk (milk/lactose intolerance and intolerance of dairy products-other than lactoglobulins, which are allergenic). Allergen reduction by thioredoxin changes the biochemical and physical properties of proteins. According to present evidence, thioredoxin may be used to improve foods through, among other changes, lowering allergenicity and increasing digestibility. Using dogs, researchers have shown that thioredoxin reduces disulfide bonds of allergens, and thereby alters the allergenic properties of proteins extracted from wheat flour.

By changing the levels of expression of the thioredoxin gene, scientists have been able to reduce the allergenic effects of the protein fractions extracted from wheat and other cereals. Thioredoxin mitigated the allergenicity associated with the major protein fractions such as the gliadins (including the alpha, beta, and gamma types) and the glutenins, but gave less consistent results with the minor fractions, the albumins and globulins. One soybean storage protein (P34) accounts for 85% of IgE responses in soybean-sensitive individuals.

Sense suppression (gene silencing), driven by a seed-specific β-conglycinin promoter, was used to eliminate the accumulation of P34 in transgenic soybeans, removing the principal source of food allergenicity in soybeans. Early results from human blood serum tests indicate that P34-specific IgE antibodies could not be detected in soybean-sensitive people fed the gene-silenced beans.

Toxins

Potatoes and tomatoes are members of the deadly nightshade family and can contain toxic glycoalkaloids (for example, solanine) that have been linked to spina bifida. Lectins are toxic glycoproteins that have the ability to bind to carbohydrate-containing molecules on the epithelial cells of the intestinal mucosa, thus causing toxicity. They are also called hemaglutinnins, based on their ability to agglutinate red blood cells.

Kidney beans contain phytohemagglutinin and are poisonous if undercooked. A number of people die each year from cyanogenic glycosides from peach and apricot seeds and many become ill from the sodium channel binding of grayanotoxin in honey produced from the nectar of rhododendrons. It is conceivable that biotechnology approaches can be employed to downregulate or even eliminate the genes involved in the metabolic pathways for the production, accumulation, and/or activation of these toxins in plants.

For example, the solanine content of potato has already been reduced substantially using an antisense approach, and efforts are underway to reduce the level of the other major potato glycoalkaloid, chaconine. Work has also been done to reduce cynaogenic glycosides in cassava through expression of the cassava enzyme hydroxynitrile lyase (HNL) in the roots.

Metabolic engineering is generally defined as the redirection of one or more enzymatic reactions to improve the production and accumulation of existing compounds, produce new compounds, or mediate the degradation of compounds. Significant progress has been made in recent years in the molecular dissection of many plant pathways and in the use of cloned genes to engineer plant metabolism. There have been numerous success stories, as well as a number of research studies that have yielded unintended results, such as attempts to modify photosynthesis.

Trait modifications with the additions of 1 or 2 genes that do not act on central or intermediary metabolism produce targeted, predictable outcomes, whereas major modifications of metabolic pathways can produce unanticipated effects. It is, therefore, very encouraging that the presently available analytical technologies have been able to detect and assess the safety of these unanticipated effects.

In addition, regulatory oversight of GM products has been designed to detect such unexpected outcomes in GM crops. As more metabolic modifications are introduced, we must continue to study plant metabolism and the interconnected cellular networks of plant metabolic pathways to increase the likelihood of predicting pleiotropic effects that may occur as a result of the introduced genetic modification.

13

Nutritionally Improved Food Feeds

There is a general agreement that the standard of safety that should be applied to food products derived from GM crops should be equivalent to that applied to foods and feeds derived through traditional plant breeding. It is a fact, however, that, unlike most foods derived from traditional plant breeding, nearly all new foods and feeds derived from GM crops have been subjected to detailed compositional analysis and many have been assessed in toxicological and nutritional studies.

So, while the standard of safety may be the same in both cases, foods derived from GM crops have been subjected to more detailed scrutiny from the point of view of safety and nutrition. In keeping with internationally recognised principles for the safety assessment of foods derived from GM crops, the general approach involves comparison of the newly developed food with a suitable comparator food that has a history of safe use. This concept, referred to as substantial equivalence, includes a detailed comparison of agronomic features and composition of key nutrients, antinutrients, and natural toxicants of the new crop compared to the conventional counterpart.

The purpose of this evaluation is to identify similarities and differences between the new variety and its comparators. Any differences then become the focus of the safety assessment. Sufficient experience has been gained with the more than 50 GM crops that have been assessed by regulatory agencies, to date, to state with considerable confidence that the process of biotechnology as applied to date has not resulted in major unintended compositional changes in the food or feed.

Indeed, as predicted, the application of biotechnology has resulted in minimal or no change in composition apart from the intended expression of specific traits. In addition, because the novel protein introduced is examined closely with respect to toxicity and allergenicity, it can be concluded that GM crops are as safe as their conventional counterparts. With this experience in hand, the challenge is to develop safety assessment procedures that can be applied to nutritionally improved GM foods and feeds.

The fundamental purpose here is to determine whether the composition of a nutritionally improved variety differs significantly from its traditional counterpart aside from the intended change in nutrient composition and to assess the safety of the intended change and any unintended changes.

Nutritionally improved varieties may be expected to contribute significant new sources of dietary nutrients or other bioactive phytochemicals. To assess the safety and nutritional impact of these products, it is important to have knowledge of how much of these products will be consumed in the overall human diet or in animal feeds. The safety and nutritional quality of these products can only be assessed in the context of their proposed uses and consequent intake.

Principles

Safety Assessment

A key basic principle is that both foods and feeds should meet the same safety and quality standards and should be subjected to the same safety assessment procedures. In the case of nutritionally improved foods and feeds, there is no single safety assessment approach that can be applied to all new products, although some core procedures, such as compositional analyses, that have been applied to GM crops to date are warranted.

The guiding principle in approaching the safety assessment is to have clear understanding of the introduced genetic changes and how these changes affect the nature and amount of expression products and metabolites. Since the types of nutritionally improved products anticipated are diverse, each new product must be approached on a case-by-case basis, applying the general principles that have evolved for products derived from GM crops with improved agronomic traits.

Exposure Assessment

Because nutritionally improved varieties may be expected to have major changes in the amounts of one or more nutrients, assessing human and animal exposure to these products is important, particularly if the exposures are significant. Exposure to altered levels of nutrients, such as fatty acids, from foods and feeds derived from GM crops needs to be considered in the context of total dietary exposure consumption of those same substances, which may appear in the diet from multiple sources. This will require knowledge of how much of the product is consumed in the diet of humans or, in the case of livestock, the extent to which it is used in animal diets.

A key consideration in the exposure assessment is the criterion that will be used to assess whether the use of a new variety will result in a significant change in dietary intake to the nutrient of interest. The word "significant" as used here refers to a change in the dietary intake of a nutrient that has the potential to materially affect health, rather than simply some defined percentage change in composition of that nutrient in the new variety. It is conceivable that a large and unintended change in content of a specific nutrient in a given food could have relatively little effect on human nutritional status with respect to that nutrient.

In contrast, seemingly small decreases in content of a specific micronutrient might conceivably have serious effects on a specific at-risk subpopulation that has marginal intake of that nutrient. The issue of what constitutes a significant change in intake of nutrients was discussed in the report of the International Food Biotechnology Council. For nutrients, it was recommended that if a food supplies less than 5% of the average daily need (intake) in an amount of the food typically consumed per day by the population in question, then the intake from that source can be regarded as nonsignificant.

Similarly, it could be stated that if the intake of any inherent constituent from a food or feed derived from a GM crop were increased by 5% or less, that would not be considered a significant change. As pointed out by IFBC, the distinction between a nonsignificant and a significant change is judgmental. The determination of the significance of a change in the level of a nutrient will also vary depending on the nutritional importance of the food and the availability of the nutrient in the food supply of the population.

Recommended dietary intakes can be or have been set for most nutrients. Since each nutrient has a unique role and function and is present at different levels in different foods, the potential impact of changes in the dietary content of nutrients must be assessed on a case-bycase basis. It should also be recognised that certain new varieties may be developed to achieve a particular nutritional purpose within a specific age or gender group. This will require that intake assessment be tailored to the specific demographic group who consume the greatest amount of the new product.

Methodologies for assessing intake of nutrients and other dietary constituents are widely available. These range from per capita methods to methods that use available food consumption databases or to actual food consumption surveys. Per capita methods include food availability estimates or food disappearance data, presumably food eaten.

Although per capita methods provide a representative general population mean of food consumption, they cannot provide consumption estimates for specific segments of the population. Specific segments may include populations who consume greater amounts of particular foods, either as a function of age, health status, or choice. Food consumption survey methods vary in their design and collection of dietary intake data and can range from 24-h dietary recalls to multiple-day dietary records. It is well known that shortterm food consumption data do not represent actual intake over a longer time period.

Twenty-four-h dietary recall data have been found to overestimate consumption of specific food components, particularly for users or eaters of specific food products. In addition, these types of surveys are generally considered to provide worst-case estimates of consumption because of the numerous conservative assumptions inherent in the methodology for estimating intake. Because of significant intraperson variability in food consumption, food consumption does not follow a normal distribution and it is difficult to determine accurately the consumption of those individuals in the 90th to 99th percentile. The greater the length of the dietary survey, the more accurate are the consumption estimates of consumers at the extremes of consumption.

Nutritional Evaluation Issues

The recommended approach for the safety and nutritional evaluation of foods derived through biotechnology involves a thorough knowledge of the parent or traditional crop, molecular characterisation of inserted DNA, evaluation of the safety of any proteins and other products expressed from the inserted DNA, application of the concept of substantial equivalence to identify similarities and differences in composition in comparison to suitable control conventional counterparts, and the evaluation of the safety and nutritional consequences of the intended alterations in nutrient composition and any other alterations identified.

Molecular Characterisation

A core component of the safety assessment of foods derived from GM crops is the molecular characterisation of the introduced DNA. A primary purpose of this analysis is to establish that the integrity of the vector DNA has not been modified as a result of the transformation process. The molecular characterisation of GM plants is comprised of essentially 2 basic components

— a comprehensive description of the genetic elements and constructs used for plant transformation, and

— the description of those elements as integrated in the transgenic event of interest.

Regulatory requirements for molecular characterisation may be different for Europe, Japan, Australia and New Zealand, Argentina, and other countries.

Transformation System and DNA

The constructs and transformation method used to generate the GM plant must be described. This includes a detailed description of the transformation method (for example, Agrobacterium-mediated transformation or direct transformation by methods such as particle bombardment, electroporation, or PEG transformation of protoplasts).

For Agrobacterium-mediated transformation, the strain designation of any Agrobacterium used during the transformation process and how the Ti plasmid based vector was disarmed should be described, as well as the process used to free the system of remaining Agrobacterium cells once transformation was complete. For direct DNA-based transformation systems, the information should include information on whether the system utilised a pathogenic organism or nucleic acid sequences from a pathogen; how such sequences, if present, were removed prior to transformation; and whether the transformation process involved the use of helper plasmids or a mixture of plasmids or carrier DNA.

A detailed physical map of the vector used for transformation, including as appropriate the location of restriction sites, should be supplied, noting those portions of the vector used as primers in PCR analysis or as probes in Southern analysis. In addition, a summary of all genetic components that comprise the vector, including coding regions and noncoding sequences of known function, should be supplied.

The data on coding regions should detail the size of the individual DNA elements; the location, order, and orientation of the elements in the vector; the source of each element; and their probable function (if any) in the plant. In addition, information indicating whether any of the donor organisms or derived genetic components are known to cause disease or injury to plants or other organisms or are known toxicants, allergens, pathogenic factors, or irritants is supplied.

If there is a history of safe use of the donor organism(s) or components thereof, that is also taken into account. With regard to coding sequences (open reading frames), significant DNA sequence alterations to the native gene that resulted in a change in the amino acid sequence must be described. If the modified amino acid sequence has not been previously published, the complete sequence (highlighting the modifications) is to be reported, while DNA sequence modifications that affect only a few amino acids can be described without providing the complete sequence. Modifications known or anticipated to result in posttranslational modifications or alterations to the structure or function of the gene product must be described.

The complete nucleotide sequence of the DNA that is transformed into the plant is not generally required. However, sufficient data must be provided to demonstrate that the nature and order of the genetic elements as they existed in the vector DNA used in the transformation process have not been substantially altered following introduction into the plant. This may include Southern blot analysis, analysis with appropriate PCR analysis, DNA sequencing, RT-PCR data, and characterisation of the protein product produced from the inserted DNA to demonstrate that the expected protein is expressed in the plant.

Data describing the number of gene copies inserted into the plant, including the integration of partial gene fragments should be provided. In the case of allopolyploid plants, information identifying which parental genome the transgenic DNA has inserted into may also be required.

The pattern and stability of inheritance of the introduced DNA (and gene function) must be demonstrated for plants that are male or female fertile, or both. A variety of methods can be used to demonstrate this, such as retention of phenotype, immunoassays, PCR, or Southern hybridisation. For plants that are infertile or for which it is difficult to produce seed, data must be provided to demonstrate that the transgenic trait is stably maintained and expressed during vegetative propagation over a number of generations appropriate for the crop.

DNA is an integral part of every plant cell and is rapidly degraded by normal digestive processes, leading a number of organisations to conclude that consumption of DNA, including DNA introduced into GM crops, is safe. To date, fragments of low-copy plant transgenes have not been detected in the tissues of animals that are typically consumed by humans.

Protein Safety Evaluation

As with foods and feeds derived from GM crops with improved agronomic traits, the safety of any protein(s) that may be expressed from the inserted DNA in nutritionally improved products derived from GM crops as a result of any genetic change must be established. The need for studies to support safety requires consideration on a case-by-case basis and depends, in part, on available knowledge about the function and biological activity of the protein, as well as any history of prior exposure.

Where appropriate, safety studies may include standard animal testing to evaluate toxicological effects or immunological studies and bioinformatic approaches necessary to assess potential allergenicity. This may require the isolation of the protein from the plant or the synthesis of the protein by other means such as by E. coli, in which case there is a need to demonstrate biochemical, structural, and functional equivalence between this test material with that found in the plant.

Substantial Equivalence Application

The concept of substantial equivalence as a starting point for the safety assessment of GM crops. A joint FAO/WHO consultation in 1996 and the Codex Alimentarius Commission of FAO/WHO in 2000 and 2002 endorsed the concept as a strong and robust starting point for the safety assessment of GM crops. Application of the substantial equivalence concept is not a safety assessment per se, but provides a basis to identify similarities and differences between the new variety and some suitable comparator variety. Differences are then subjected to further safety assessment.

Nutritionally improved products are expected to consist of 2 categories of products. One category will be nutritionally improved foods and feeds intended to replace traditional varieties in the human diet or in animal diets. The 2nd category of products is food or feed ingredients derived from nutritionally improved crops. Some of these will be identical chemically to ingredients currently derived from food crops, whereas others could be chemically altered products, such as cross-linked modified starches that are modified to have specific processing or health attributes.

Compositional analysis: Compositional analysis is the major factor assessed in the determination of substantial equivalence. Various grain, plant parts, and/or processed fractions are analysed to determine the amounts of specific analytes in the matrix. These analyses range from the crude proximates (protein, fiber, fat) to very detailed analysis of the amino acid composition of the matrix. Thus a typical composition profile consists of moisture, crude protein, crude fat, ash, fiber fractions, amino acid and fatty acid profiles, vitamins, and minerals. In addition, data on antinutrients and other biologically significant compounds present in the crop, such as trypsin inhibitors, endogenous toxins, isoflavones, or phytic acid should be obtained.

It is critical that data used in the assessment of composition are statistically robust. This means that the data must come from a sampling plan that has been set up to a defined protocol in order to obtain a representative and substantially robust sample. Developers have often adopted practices employed in pesticide residue trials, as required by EPA and in line with Codex recommendations.

In other studies, replicate samples are collected or samples are collected from multiple plots at the same location. In some cases, the sample may be from a much larger number of plants (for example, from a bulk sample from a large plot), and in these cases care must be taken to obtain a representative sample from the bulk sample, either by employing appropriate sampling methods or by sampling multiple times while harvesting the plot. Although many of the analytes show a normal distribution, this cannot be assumed.

Thus, a statistical test that is relatively insensitive to such effects is best utilised. When comparing data, care must be taken to account for the distribution of the data.

Selection of appropriate comparator: One of the key considerations in applying the concept of substantial equivalence is the selection of an appropriate comparator. If a specific food or feed component is modified (for example, the fatty acid content of the oil), it may be more appropriate to compare the component to the composition of the oil from another crop or other source than to the oil from the crop that was modified.

This method was used for canola with increased levels of lau-rate, in which the oil content was compared with tropical oils instead of with conventional canola oil. Two approaches are in use. In the first approach, the package should include data from a genetically similar comparator grown alongside the GM crop as well as data on the range of composition from other varieties of that crop (data specifically generated or from the published literature).

In some cases, the GM crop has also been compared to a number of commercial varieties. In practical terms, applicants wishing to register GM crops have carried out both comparisons. There are a number of limitations to this approach. The first is that, although a comparator may be considered near-isogenic, it is certainly the case that normal Mendelian genetics result in a large number of genetic loci potentially differing between the GM crop and the closest comparator. This is especially true where the comparator is not a line that has been specifically bred to be a comparator for the line being tested.

In the second approach, the data obtained from the GM crop are compared to the publicly available data. For maize, data are typically obtained from publications that have been compiled for the feed trade. The biggest concerns about these data are that the sources are often dated and lack association with specific analytical methods. Users therefore cannot compare their data directly with data obtained using the same quantitative methods.

To alleviate this problem, the ILSI International Food Biotechnology Committee constructed a comprehensive upto-date database on the composition of crops that is accessible via the internet. By pooling data generated by the agricultural biotechnology industry, the scientific basis for comparison of composition data with the larger data set of each crop will be significantly improved. Public data that meet the acceptability criteria will be accepted added to the database, so that other publicly available data can be incorporated in a consistent manner from throughout the world.

This robust database will further the understanding of the phenotypic diversity in composition of conventional crops and their products and will allow better evaluation of the composition of nutritionally improved GM crops and their products.

The recommended approach for the safety and nutritional evaluation of nutritionally improved foods and feeds follows concepts already successfully employed for the evaluation of products derived from GM crops with improved agronomic traits. As indicated previously, foods and feeds derived from GM crops with improved agronomic traits have not been reported to be significantly altered in terms of the concentrations of macro- and micronutrients and other inherent constituents, providing a high degree of confidence that the amount of food and feed from nutritionally improved GM crops will not present new safety issues.

Hence the safety and nutritional assessment of these products can rely on historical practices employed to date. The range of new nutritionally improved products derived from GM crops is potentially very diverse, including varieties with altered levels of amino acids (for example, high-lysine maize) and vitamins, reduced levels of antinutrients (for example, phytates), altered fatty acid composition, and the use of plants for the production of new ingredients that may not be native to the plant.

In approaching the evaluation of the safety and nutritional value of such products, 2 key questions emerge. The first of these is how the product will be used. Is the product intended to be consumed as a whole food or feed replacing a traditional product, or is it intended that the product of the genetic modification will be separated from its plant production system and consumed as an ingredient? The approach to safety assessment will be different in these 2 cases. The second key question that emerges relates to the extent of consumption of the new nutritionally improved food or ingredient. This must be known or predictable in advance of performing a safety or nutritional evaluation.

Nutritionally improved foods or feeds derived from plants and intended for use as replacements for traditional products are best compared initially with their parental varieties. The initial approach is to apply the concept of substantial equivalence focusing on constituents other than the altered level of nutrients. Detailed analysis of major and minor constituents should be undertaken with a view to determining whether the intended genetic change has altered the concentration of inherent constituents other than the intended improvement in nutrient composition.

If no significant changes are observed from compositional analysis, the safety and nutritional evaluation then focuses on the altered levels of nutrients arising from the genetic modification. It should be established that, under the conditions of intended use of the new food or feed, there is no increased safety concern due to the altered level of nutrients compared to the traditional source. As noted above, a key dimension of this is determining the most likely exposure level for the altered nutrient(s).

Safety can only be evaluated in the context of use patterns and exposure. For new crops that contain altered amounts of nutrients, the range of safe intakes can be established from the literature. For example, there are adequate data on amino acid or fatty acid toxicity to establish whether altered concentrations of these substances in a whole food/feed would present a safety concern. It can be concluded that, for the vast majority of new nutritionally improved GM varieties, the principal focus will be on enhancing nutrient composition or improving bioavailability or functionality of existing inherent constituents.

Such compositional changes are unlikely to raise safety concerns because of the well-established role of nutrients in human and animal nutrition. The only residual issue of potential concern might be the presence of unintended changes in composition or metabolic pathways. In cases where the nutrient is separated from its plant source with the intention to use it as an ingredient in foods or feeds, the use pattern and exposure again dictate the approach to the safety assessment.

Information must be obtained on how the product will be used and the consumption that might be anticipated from its use. As indicated earlier, nutrients derived from nutritionally improved crops may be chemically identical to existing nutrients or they may be chemically altered to improve their functional or physiological properties. The use of these materials in food or feed will be subject to existing regulations, and chemically altered substances may require detailed safety assessment and regulatory approval prior to use.

Role of animal tests: Historically, toxicity tests in laboratory animals have played a significant role in ensuring the safety of chemicals present in foods, including food additives and contaminants that typically are consumed by humans in very small amounts. However, their value for assessing the safety of whole foods or major food constituents presents a number of difficulties. Before considering this matter, it is important to point out that, consistent with the concept of substantial equivalence, the safety assessment of foods derived from GM crops focuses on the examination of any differences between a suitable traditional variety and the new GM variety.

This concept also holds in the conduct of animal tests where test groups are fed the food derived from the GM crop while the control group is fed a suitable comparator food. A key challenge for future consideration is the role of animal tests in the safety assessment of new GM varieties with significantly different nutrient composition from traditional varieties. In these cases, suitable comparator (control) varieties may not be available and existing study protocols may need revision to ensure the safety assessment is appropriate and adequate.

The difficulties encountered in assessing the safety of foods derived from GM crops in bioassays such as animal tests are well recognised. It has been pointed out on numerous occasions that animal feeding studies with whole foods or feeds must be designed and conducted with great care to avoid problems encountered with nutritional im-balance from overfeeding a single whole food, which itself can lead to adverse effects.

In undertaking such tests, a balance must be struck between feeding enough of the test material to have the possibility of detecting a true adverse effect and, on the other hand, not inducing nutritional imbalance. In any event, the multiples over anticipated human intake one would like to attain in animal tests are simply not achievable for practical reasons, and margins of safety of 1 to 3 times have to be accepted. This limits the sensitivity of animal bioassays to detect small differences in composition, which may be more readily detected with thorough analytical characterisation.

Even though animal tests lack the sensitivity to detect minor changes in composition, in some instances, properly designed studies can confirm conclusions from other elements of the safety assessment and provide added assurance of safety. However, it must be recognised that the ability of rodent bioassays to detect adverse effects from an inherent constituent of a food derived from a GM crop depends upon the intrinsic toxicity of the constituent and whether it is present in the food in sufficient amounts to induce toxicity under conditions of a bioassay.

In general, it is difficult to feed experimental animals more than 25 to 30% of the diet of a food product without creating nutritional imbalances, so the concentration of toxicant would have to be sufficiently high (or the toxicity so significant) in the food product portion of the rodent diet to produce toxicity. If it is not, the rodent bioassay simply will not detect the presence of the toxicant.

The broiler chicken has emerged as a useful animal model for assessing nutritional value of foods and feeds derived from GM crops. It should be noted, however, that, contrary to laboratory rodents, the rapidly growing broiler has been obtained through breeding efforts with the aim to create an efficient food-producing animal. This may, therefore, not render it optimal for toxicological testing of foods and feeds. In fact, disorders such as "sudden death syndrome" and "ascites," are considered related to metabolic disorders associated with its rapid growth.

On the other hand, broiler chickens have been optimised for growth relative to highly characterised diets such that small changes in nutrients or antinutrients in the diet are readily manifested in reduced growth. In addition, one of the first indications of an ill animal is loss of appetite or reduced growth rate. Also associated with the rapid growth of broiler chickens is the reduced fertility of overweight broilers allowed ad libitum access to feed.

Live weight gain, efficiency of feed conversion, carcass weight, and breast muscle and fat pad weight are the traits usually measured in broiler feeding studies with feedstuffs from GM crops. Given the background of adverse symptoms related to the rapid growth of these animals, it seems that broiler chickens are not as useful for toxicological testing as are the common laboratory animals such as rats, mice, rabbits, and guinea pigs.

Among the traits measured are body and organ weight, feed consumption and conversion, blood chemistry, serum IgE and IgG levels, urine composition, hepatic enzyme activities, and histopathology of organs and intestinal tissues. There are no indications from these experiments that unintended effects that affect animal health or productivity occur as a result of the genetic modification process.

The key issues to be considered here are the formulation of diets with appropriate nutritional characterisation and the avoidance of diets that are nutritionally unbalanced. A further issue is the selection of an appropriate control diet. Ideally, the control diet should be comprised of the foods or feeds selected for the analytical trials.

Improved nutrition products derived from GM crops may differ considerably in nutrient composition from traditional varieties making direct comparisons difficult. It is also essential that the experiment is properly designed with an adequate number of replications to provide sufficient statistical power. Clearly, each new food or feed derived from a GM crop needs to be assessed on a case-by-case basis and it is not possible to formulate in advance any routine approach to animal safety testing.

14

Nutritional Assessment Process for Biotechnology

An important part of the safety assessment of nutritionally improved GM crops is the need to develop effective methods to evaluate the potential impact of such crops on the diet. Such analyses are complex for a number of reasons, among which are:

— the challenges associated with estimating dietary intake of specific foods,

— the highly variable nature of the human diet,

— the complex relationship between individual nutrients and human health,

— the possibility of nutrient-nutrient interactions, and

— the unique lifestyles and genetic makeup of individuals.

The nutritional impact of a few major plant-based commodities such as wheat, rice, potatoes, and maize is relatively easily assessed since they may comprise a large and relatively fixed proportion of the human diet for some populations. The impact of a significant change in the content of a nutrient, for example β-carotene in "Golden

Rice" or "Golden Mustard," can be reasonably simple to compute. Changes in foods that comprise a large percentage of the dietary intake also have good potential for significant nutritional impact. In many industrialised countries, the average consumer's diet can be derived from hundreds or even thousands of food products.

Dietary intake patterns may .also change with time. New products and new consumer preferences for these products may gradually change the overall diet. Reliable methods for establishing the current intake of nutrients of a population and estimating the range of changes in nutrient intake that might be caused by a nutritionally improved product are needed to assess its impact on nutrition and health.

One or more of the following types of changes could be incorporated into nutritionally improved GM crops:

- Increases in the content of a single nutrient in a manner such that there is no significant impact on other components but there is an enhancement of nutritional value (for example, enhancing the lysine content of maize or enhancing the level of β-carotene of "Golden Rice" or "Golden Mustard").
- Increase in the content of a nutrient with a compensatory decrease in another (for example, increasing the protein content of a grain with a compensatory decrease in carbohydrate or lipid content, as occurred in Quality Protein Maize [QPM].
- Alteration in the composition of a nutrient with no concomitant change in overall macronutrient composition (for example, exchange of linolenic acid and linoleic acid by oleic acid to make a high-oleic acid vegetable oil).
- Changes in bioactive substances, such as an increase in an antioxidant, with expected benefits to health.

— Reductions in the content of an antinutrient, toxin, or allergen.

— Changes that affect bioavailability, absorption, or utilisation of a nutrient (for example, low phytate or high phytase content in soybeans as a means to enhance phosphorus or trace mineral absorption).

Changes in macro- or micronutrient content intended to improve quality traits can impact nutritional value as well. Increased starch content in potatoes and elevated solids content in tomatoes are examples of changes that improve nutritional value and impact functional qualities as well. High-oleic vegetable oils were first developed as stable frying oils, but were later recognised for their potential to influence dietary fatty acid intake. It is clear that alterations in composition that might change dietary intake of nutrients should be evaluated regardless of the intended purpose for which they were developed.

Nutritionally Improved Foods

There are at least 2 primary ways in which nutritionally improved foods might be introduced into to the consumer's diet. Foods and food ingredients that have enhanced nutrient content and/or bioavailability will be developed and marketed to consumers as higher value products that identify the enhancement on the label. It would be required to state changes in nutrient composition on the label, and the developer would almost certainly want to market some perceived benefit of the product's consumption to consumers.

It is important to point out that nutritionally improved food products that are described in labeling or advertising as having particular health benefits (for example, cholesterol reduction) will require the generation of data to substantiate these claims. These data are distinct from the data that are generated to demonstrate the safety and nutritional value.

The alternative pathway by which an enhanced nutrition product might reach consumers is as a food or food ingredient intended to replace a generic commodity. An example might be the introduction of higher levels of sulfur- containing amino acids into soybeans, or the increase in lysine and tryptophan concentrations in maize. It is conceivable that a thorough safety review would provide evidence that these enhanced commodities were in all ways as safe as their traditional counterparts, and that they provided improved essential amino acid intake.

Thus, setting aside commercial and market considerations, one could conceive of approving the wholesale replacement of the traditional varieties of maize and soybeans with these new nutritionally improved varieties. Such wholesale replacement would be analogous to food fortification, and this possibility is in fact sometimes called biofortification. Same scientific logic that applies to national policies regarding nutrient fortification could apply to these novel commodities. From a scientific point of view, the process used to create the fortification is immaterial to the evaluation of safety and nutritional efficacy.

Quality Protein Maize (QPM) describes a family of maize varieties that contain higher levels of lysine and tryptophan than conventional maize. In addition, QPM has twice the protein content of conventional maize and is 10% more productive in the field relative to local varieties. Quality protein maize was developed to improve the protein nutrition of populations who consume significant quantities of maize in the diet and are at risk of protein malnutrition.

The initial development of QPM was the isolation of the high-lysine maize mutant opaque-2. High-lysine maize that would improve the essential amino acid balance of maize for use in animal feed has also been under development by plant breeders for over 40 y.

Nutritional Safety Assessment of Nutritionally Improved Crops

It is worth reiterating that the process described herein to evaluate the safety and nutritional value of nutritionally improved products derived via biotechnology is basically no different than the process used to evaluate the nutritionally improved products developed by any other technology (for example, traditional breeding, use of novel food ingredients). The analysis does not differ significantly in principle from that applied to new food ingredients. The sequence of steps in the analysis for such nutritionally modified foods is as follows:

- — Estimation of magnitude of changes in nutrients (compositional analysis)
- — Determination of the expected level of use in food products
- — Estimation of frequency of use in products (foods in which the new variety or product is incorporated or for which it is substituted; generation of food lists)
- — Estimation of anticipated distribution of dietary intakes by selected groups (intake ranges by age, gender, demographic, health status), also known as "exposure assessment"
- — Assessment of the potential nutritional and health outcomes, taking into account nutrient content and bioavailability and the effects of processing.

The strategy for the evaluation of the nutritional impact for a crop variety that has altered nutrient composition can also be applied to the evaluation of changes in nonnutritive food components that are intended to provide health benefits. Specifically, plant breeders are developing crops with elevated levels of vitamin C, vitamin E, and β-carotene, nutrients with well-known safety and nutritional benefits.

Vegetable Oils with Altered Fatty Acid

There are several examples of vegetable oils in which fatty acid compositions have been altered to produce nutritional and/or functionality improvements. It is instructive to review the safety assessment process undertaken for each product as a way to illustrate the process that could be established for other nutritionally improved products.

— High-oleic soybean oil was first developed by traditional plant breeding methods to improve the oxidative and thermal stability of soybean oil by replacement of polyunsaturated fatty acids with oleic acid. Varieties have also been developed using biotechnology. In these varieties, the formation of monounsaturated oleic acid is favored over the production of polyunsaturated fatty acids due to suppression of Δ-12 desaturase activity by the inserted DNA. In addition, the allergenicity of the high-oleic soybeans has been compared to that of conventionally produced soybeans and shown to have a similar allergenicity profile.

— High-laurate canola has been produced by insertion of a fatty acyl thioesterase gene from California bay tree. The thioesterase preferentially cleaves lauroyl groups, leading to an enrichment of lauric acid levels compared to conventional canola. This canola has completed regulatory review for food use in USA and Canada.

— Solin oil is a form of linseed oil isolated from a flax variety that produces reduced levels of linolenic acid, making it more suitable for certain processed food applications. Solin oil was developed and approved in Canada and has been self-affirmed as "generally recognised as safe" (GRAS), and the United States Food and Drug Administration has not objected.

— High-oleic acid sunflower oil was commercialised in the mid 1980s.

The potential health consequences of the replacement of cooking and frying oils with high-oleic acid soybean oil were considered in the premarket safety review. It was noted that the oil was similar in composition to olive oil. The impact of substitution of the modified oil on overall dietary intake of fatty acids was calculated based on consumption data for British consumers. Of particular interest was the estimation of the impact of this substitution on consumers at the highest extreme of consumption whose overall intake of fatty acids would be the most altered.

It was concluded that a potential "worst case" decrease of 29% in the consumption of linoleic acid would have negligible effects on the incidence of cardiovascular disease and that the concomitant reduction in saturated fatty acid intake was likely to have a more significant beneficial effect on health. It also was concluded that the intake of monounsaturated fatty acids would increase, at the expense of saturated fatty acids and n-6 polyunsaturated fatty acids.

Thus, the substitution in the diet of oil from soybeans modified to be high in oleic acid would have an effect on the diet comparable to that of substitution with olive oil. Some nutritionists have supported such a dietary change as a way to achieve the health benefits of a more "Mediterranean" diet. For the high-laurate canola, no nutritional concerns arising from its replacement in the diet of other high-lauric acid oils such as coconut oil were raised. The levels of antinutrients were also similar to those found in conventional canola.

In the GRAS affirmation of Solin oil, its dietary impact was calculated assuming it would completely substitute for sunflower oil. In reviewing Solin, of which substantial amounts are consumed in the European Union, the British Advisory Committee on Novel Foods and Processes (ACNFP) expressed in its Annual Report a concern over the dietary impact of the use of modified oils with longer shelf life:

Irrespective of whether food ingredients from this variety of linseed are considered to be novel or not, the ACNFP was concerned that the levels of α-linolenic acid had been reduced in the oil in order to improve shelf life. In reducing the levels of this fatty acid to prevent rancidity, the ratios of n-3 to n-6 fatty acids were dramatically changed in the oil. The ACNFP was concerned that altering the fatty acid ratio in this way may have long-term effects on public health. The Committee acknowledged that this problem was not unique and that there is a growing trend in altering the fatty acid composition of vegetable oils, through the use of traditional plant breeding techniques, in order to improve their shelf life but at the expense of nutritionally beneficial fatty acids.

It should be emphasised that the focus of the ACNFP was on the impact on diet and health of changes in oil composition rather than the process used in plant breeding (traditional plant breeding versus biotechnology). It should also be noted that the Canadian authorities treat plants that have been modified through biotechnology and mutation breeding as "plants with novel traits"-they do not distinguish between the process of production of the new varieties. Conventionally bred crop varieties with modified oil compositions have, therefore, been subjected to the same approval procedure as those developed with the use of biotechnology in Canada.

Varieties of high-oleic acid canola and low-linolenic soybean produced through traditional breeding have been approved for food use in Canada. In these cases, toxicity testing was not considered necessary, because of the similarity of these oils to the oils from other conventional foods and the removal from the oil of proteins that might have caused concerns about their allergenicity.

In everyday life, is not likely that sunflower oil will substitute 100% for olive oil, and number of sources will contribute to total oil intake, thus such differences would likely not be observed. The comparison of olive oil and high-oleic sunflower oil demonstrates that an adequate nutritional assessment cannot focus solely on the change in composition of a single macrocomponent of a food. While the 2 oils compared are similar in oleic acid content, significant differences in fatty acid content, triglyceride structure, and the content of other constituents gave rise to differences in the physiological effect of these 2 oils in animal and human studies.

In general, nutritional assessment should consider changes in content of all components of a food that are known to play a significant role in diet or health through consumption of that food. Which nutrients should be analysed will depend on whether that food is a significant dietary source of that nutrient, thus the assessment must be made on a case-by-case basis.

Changing Nutritional Composition

Changes in nutritional content can be quantified by compositional analysis. When a plant variety has been purposefully developed to contain altered levels of a specific nutrient or nutrients, the intended changes in those specific nutrients should be well documented. It is possible, however, that a significant compositional change in one nutrient might result in additional changes in composition. Therefore, an analysis of macro- and micronutrient composition should be performed either to document that no ancillary changes in nutrient content have occurred or to identify those changes. A list of analytes for a detailed nutrient evaluation is given below:

— Proximate analysis (protein, lipid, carbohydrate, fiber, ash, moisture)

— Amino acids

— Fatty acids

— Fat-soluble vitamins

— Water-soluble vitamins

— Minerals for which a need in human nutrition has been established

— Known beneficial nonnutritive substances

— Known antinutrients

— Toxicants

Determining the significance of a change in nutrient content of a food It is challenging to determine a mean concentration value for a specific nutrient in a given food crop such as maize, wheat, or soybeans. Large variations in content of specific nutrients are commonly encountered in different samples of the same food.

Different varieties of the same crop plant have been observed to have significant differences in composition. Representative compositional analyses of the same variety will also vary because of environmental effects such as geography, soil, climate, harvesting, and postharvest handling. In fact, several-fold variations in content of some less stable nutrients such as vitamin C are commonly encountered.

Variability in the Human Diet

The highly variable nature of the human diet and the resulting variability in nutritional states must be considered in order to fully understand the nutritional adequacy of a food. Data on human dietary intake are usually gathered through carefully designed food intake surveys. Intake studies present a number of methodological challenges, not the least of which is the need to rely on self-reporting of intakes by subjects.

The most useful studies survey large, demographically representative populations repeatedly over a number of years, while simultaneously tracking an array of environmental, social, and health parameters. Protocols have improved markedly in recent years.

Nonetheless, reliable comprehensive dietary intake data are only available for a few countries such as the United States and the United Kingdom. The United Nations recently has placed emphasis on defining nutritional status and health around the globe, with particular attention to food insecurity and the definition of "at-risk" populations. The most critical nutrient deficiencies in regions and subpopulations of most countries are now fairly well documented.

Insufficient energy intake is often accompanied by a shortfall in specific nutrients as well; in some cases, energy intake is adequate while undernutrition of specific nutrient(s) is prevalent. Worldwide, vitamin A, iron, iodine, and protein deficiencies dominate the list of nutrients for which numerous subpopulations are at risk for undernutrition.

Assessment of the potential health impact of a change in a single nutrient is a prerequisite for evaluating the potential impact of the new variety on the adequacy of the diet, but it is not sufficient information. The role of folate in masking vitamin B_{12} deficiency is a classic example that illustrates the need to evaluate the diet rather than the intake of an individual nutrient. As noted previously, the health impact of all the changes in nutrient composition in an individual new variety should be assessed-if these changes are deemed "significant."

A comprehensive analysis should include the change in composition multiplied by the percent each food containing the novel variety represents in the diet. Each of these contributions is then summed to determine the overall dietary intake. This process needs to be repeated for all nutrients so the potential health impact of a change in composition of a single nutrient can be evaluated in the context of intake of all nutrients. Some novel foods will be altered with respect to specific components such as fatty acids, amino acids, or specific vitamins.

If no collateral changes in nutrient composition are present in the variety, such changes can be viewed as a means to augment the intake of specific nutrients. The substitution of a monounsaturated fatty acid for a saturated fatty acid or an essential amino acid for a nonessential one might improve nutritional quality.

If it is proposed to substitute a nutritionally enhanced crop for a conventional crop as one means of biofortification for a specific nutrient, it should be shown that the change in composition results in an improvement in nutritional status of the population for that nutrient. As with other fortified foods, on a case-by-case basis, digestibility and absorption studies may be appropriate to demonstrate that the enhanced nutrient is bioavailable.

Any potential antagonisms, for example, competition in amino acid uptake, should also be assessed. Finally, it should be demonstrated that the increased intake of the novel food does not simply replace another source of the same nutrient or partially displace another nutrient from the diet.

15

Application of Modern Biotechnology

Biotechnology-derived food plants with improved nutritional properties are under development and, in a number of cases, have reached the field trial phase and reached commercialisation. This new generation of food/feed plants has the potential to

- combat nutrient deficiencies in the population through the introduction of essential nutrients,
- improve the nutritional value of foods/feeds by enhancing essential nutrients or by improving their bioavailability,
- promote human health through elevated levels of health protecting/promoting compounds, and
- improve the safety of products by lowering levels of natural toxins, toxic metabolites, or allergens.

In addition to human nutrition, functional characteristics are being developed that are targeted at animal feed and food processing. Multiple gene inserts coding for complex components of metabolic pathways have been engineered in some of these plants.

A key question to be addressed is whether these types of food and feed products require additional safety testing compared to the "first generation" of GM plants (those with improved agronomic properties such as insect protection, herbicide tolerance, or a combination of these traits) because of a greater chance of unintended changes in composition.

The safety of conventional crops is generally based on a history of safe use. Formal safety assessment is not conducted. For new varieties, analysis of agronomic performance and other phenotypic characteristics, as well as of selected macro—and micronutrients, antinutrients and toxicants, is performed. This has given us a wealth of data on the presence of nutritionally beneficial compounds as well as antinutrients and toxicants, whose levels have been either increased and/or diminished by the selective procedures.

Plant breeders have discarded from the breeding programmes products with unusual agronomic performance, unpleasant taste, or harmful levels of a specific compound. These approaches toward selection of conventionally bred crops have proven their efficacy and have provided the consumer with safe and nutritious foods.

Similar, but much more extensive, practices are applied to foods derived from GM crops to assess their safety and nutritional equivalence to conventional crop varieties. New analytical technologies, based on genome research and modern chemical-analytical developments, have emerged over the last decade. These technologies may provide interesting information about the structure and function of the genome of plants. These methods also give more insight into the physiology and metabolic pathways in plants, including the regulation of levels of beneficial and toxic constituents.

The potential occurrence of unintended changes in foods produced through the application of recombinant DNA (rDNA) technology is a topic that has attracted broad interest from scientists, regulators, and consumer groups. Insertion of DNA sequences into the plant genome may lead to modification of gene function, possibly resulting in shifts in metabolic pathways, upstream and downstream effects, changed metabolic pools, and formation of new metabolites or changes in levels of existing metabolites.

Identification of unintended effects in GM crops and their assessment with respect to human and animal safety has been one of the topics of the European Network Safety Assessment of Genetically Modified Food Crops, sponsored by the European Commission, and has recently been reviewed. Some unintended effects may be predictable on the basis of what is known about the site of insertion of the introduced DNA and function of potentially disrupted gene(s) and/or the function of the inserted trait and its involvement in metabolic pathways. Other effects are less predictable because of the limited knowledge of gene regulation and gene-gene interactions.

Unintended Effects in GM Crops

The occurrence of unintended effects is not specific to genome modification through recombinant DNA technology, as unintended effects also occur frequently as a result of conventional breeding and during the use of chemical and irradiation mutagenesis to produce new phenotypes. Insertion in plant DNA must, therefore, be evaluated with an awareness of such natural DNA events, an issue that has been reviewed recently. The molecular changes that occur during conventional breeding may nevertheless be less easy to trace than changes caused by insertion, rearrangement, or mutagenesis of exogenous, distinguishable DNA originating from the same or other species.

Natural chromosomal recombination mechanisms play an important role in plant breeding, and mechanisms can be grouped into

— homologous recombination and

— illegitimate recombination, a nonhomologous end-joining process.

Both processes are characterised by double-strand break repair mechanisms. Nonhomologous recombination is the predominant form in plants. Although recombination events could theoretically occur at random along the length of the chromosomes, the presence of preferential sites for recombination breakpoints is well known. Using modern biotechnology, exogenous DNA is integrated into plant genomic DNA through various methods including biolistic or microprojectile bombardment, Agrobacterium-mediated transformation, or other methods.

In the case of Agrobacterium-mediated transformation, DNA integration from the Agrobacterium Ti (tumor inducing) plasmid into the plant genome occurs in the absence of any homology with plant DNA sequences, through the process of illegitimate recombination. Insertion of introduced DNA into chromosomal DNA by this method can result in single or repeated copies and in multiple insertions.

Moreover, rearrangements of the inserted DNA and/or of the target site DNA can be observed. Sequencing the flanking regions of DNA insertions in tobacco has highlighted the presence of motifs that flank the inserted DNA, such as AT-rich sequences, microsatellite sequences, retro-elements, or tandem repeats. In Arabidopsis, AT-rich regions have been proposed to be the preferred targets for DNA introduced by particle bombardment, suggesting that the recombination processes of DNA insertion by Agrobacterium and of DNA delivered by the biolistic method are controlled by similar principles.

Based on the knowledge about DNA recombination mechanisms discussed above, there is no reason to suppose that integration of introduced DNA into plant chromosomes is more likely to give rise to DNA disruption than natural recombination mechanisms.

Identify Unintended Effects

Genome Analysis

In the situation where genomes are well mapped and much or all the genome has been sequenced, localisation and characterisation of the site(s) of insertion may be helpful to predict effects due to the DNA insertion in the recipient plant. Data on plant genome sequences flanking the inserted DNA provide information on whether the inserted DNA is within or in the proximity of a known endogenous gene.

The chromosomal location of the inserted DNA can be detected by various methods such as genomic in situ hybridisation, fluorescence in situ hybridisation, and direct sequencing of flanking DNA. Knowledge of genomes in most plants is still limited, as is the reliability of annotations in genomic databases, but the understanding of the genomic code and the regulation of gene expression in relation to the networks of metabolic activity is expanding rapidly.

Therefore, sequencing of the site(s) of insertion(s) may become increasingly informative with respect to possible alterations in metabolic networks by the modification of native DNA that may affect the toxicological or nutritional status of the modified product. However, even when the genome sequence of the host crop has been sequenced, these analyses may be more complicated than initially envisioned. Plants contain pseudogenes (for example, nonfunctional genes that are highly homologous to functional genes) and large numbers of highly repetitive

DNA sequences. These sequences make it very difficult to assess whether DNA has inserted into a functional gene, nonfunctional pseudogene or into highly repetitive DNA.

Compositional Analysis

A comparative analysis of (1) the agronomical/morphological characteristics, (2) macro- and micronutrient composition and content of important antinutrients and toxicants, and (3) the toxicological and nutritional characteristics of the modified product and its conventional counterpart tested in appropriate animal models may help to define or assess the importance of the occurrence of unintended effects.

Such an approach has been the basis for conventionally bred products with significant nutritional modification and will be explored here for its suitability for GM crops with nutritionally improved traits. A number of different strategies can be followed. Targeted approaches are hypothesis driven and focused on the generation of information on specific known macro- and micronutrients, toxic, allergenic, or bioactive compounds present in the GM food crop and its corresponding conventional counterpart.

The spectrum of compounds is based on experience gained from analysis of food crops obtained via conventional breeding. Much as in the case of human blood sample testing, analytical measurement of key substances can indicate underlying changes in health/metabolism, which can then be followed up case by case. The OECD has recently developed international consensus documents on the particular components that could be analysed for specific crops.

For the analyses of these compounds, a number of validated methods are available that are well understood through their long history of use. Such approaches work well to assess the concentrations of specific predetermined compounds that, because of their selective nature, may miss other unintended changes. By selecting the nutritionally and antinutritionally relevant compounds, this provides considerable protection against unintentional changes in known nutrients and antinutrients. However, the concentrations of preselected compounds, for example, are not necessarily indicative of changes in function or phenotype of a plant as a result of the genetic modification.

In addition, because only known substances are measured, any potential changes in the concentrations of unknown toxicants and antinutrients (if they exist) caused by the genetic modification would remain unevaluated by compositional analysis. However, it is likely that those of importance will have already been identified by their effect on health and nutrition during a history of use by a substantial portion of the consumer population, so the gap is expected to be very limited. The proposed lists of nutrients, antinutrients, and toxicants include those for which data suggest that changes in the concentrations could affect safety or nutritional quality.

Another limitation is that sample processing (for example, extraction) before single-compound assays may be so rigorous that relevant compounds may be lost before sample analysis. These limitations may be of particular importance for food plants with limited or no history of (safe) use, and for plants that have been modified extensively through multiple gene insertions. Given the fact that the GM crops currently on the market do not fall into either of these categories, single compound analysis has been sufficiently rigorous to screen for unintended effects and thereby corroborate the safety of these crops.

In the future, single compound analysis will continue to be the method of choice, using validated and quantitative methods. Nontargeted approaches for the detection of unintended effects rely on profiling methods that may provide information on potential changes in the physiology of the modified host organism at different cellular integration levels (that is, at the level of mRNA expression and protein translation and at the level of plant metabolism).

New techniques such as the DNA/RNA microarray technology, proteomics, and hyphenated (that is, coupled) analytical methods enable an integrated and simultaneous analysis of gene expression and protein and metabolite formation. Multiple factors determine the morphologic, agronomic, and physiologic properties of a food crop, such as genetic characteristics, agronomic attributes, environmental influences, plant-microbe interactions developmental stage, and postharvest effects.

Application of the new analytical technologies during various developmental stages of the plant and under different environmental conditions may provide important information on the dynamics of gene expression and resulting metabolic consequences. Profiling techniques provide an open-ended broad view of the complex metabolic networks of the organism without knowledge of changes at the level of single cell constituents (for example, organelles).

However, appropriate use of these new technologies requires validated methods with well-understood parameters and limitations, and the generation of extensive databases on conventionally bred crop varieties grown under a variety of environmental conditions in order to assess the level of natural variability. The amounts of data on baseline concentrations of natural constituents in traditional plants are, in general, limited.

In addition, the massive amounts of data generated from these methods present the challenges of proper (multivariate) statistical analysis and interpretation of changes in expression patterns, including the biological significance.

Methods of Unintended Detection

Specific Compound Analysis

Current comparative compositional analysis of GM crops and their counterparts is targeted at specific compounds, including toxicologically and nutritionally relevant macronutrients, micronutrients, antinutrients, toxins, allergens, and bioactive substances. Differences in compositional profiles that may be found to be due to the targeted genetic modification will need to be assessed with respect to functionality, toxicity, efficacy, bioavailability, and so on.

Significant changes in other nontargeted metabolic pathways leading to greater concentrations of toxic plant substances, such as glycoalkaloids in potatoes and tomatoes, will also need to be investigated further, if the changes lead to concentrations that fall outside the natural range of variability. It should be clear that, even in those cases where differences in composition between the modified crop and its counterpart are observed that fall outside these ranges of variability, such crops do not necessarily pose a threat to human or animal health.

Such differences should be assessed on a case-by-case basis as to whether additional investigations are appropriate to address further possible concerns related to the food and feed safety of the crop plant. Information on compositional analysis of GM crops in the framework of the assessment of substantial equivalence is provided by consensus documents developed by the OECD Task Force on the Safety of Novel Foods and Feed.

The ILSI International Food Biotechnology Committee has developed a database on crop composition that provides a detailed assessment of the baseline of compositional variations for conventional crop varieties, beginning with maize and soybeans.

Gene Expression Analysis

Examples of profiling techniques that could be used for compositional analysis are depicted in Figure 1.

A powerful tool to study gene expression is DNA microarray technology. The study of gene expression using microarray technology is based on hybridisation of mRNA to a high-density array of immobilised target DNA sequences, each corresponding to a specific gene. The mRNAs from samples to be analysed are labeled by incorporation of a fluorescent dye and subsequently hybridised to the array. The fluorescence at each spot on the array is a quantitative measure corresponding to the expression level of the particular gene.

The major advantage of the DNA microarray technology over conventional gene profiling techniques (for example, restricted fragment length polymorphism—polymerase chain reaction [RFLP-PCR] followed by electrophoresis) is that it allows small-scale analysis of expression of a large number of genes at the same time in a sensitive and relative (that is, within a test) manner.

Furthermore, it allows comparison of gene expression profiles of GM crops and conventional lines under different environmental conditions. This technology and the related field of bioinformatics are still in development and further improvements can be anticipated and will be necessary for use of these methods for GM crops. Current limitations to this technology are the need for microarray standards that will facilitate exchangeability and comparability of gene expression profiles of food and feed crops.

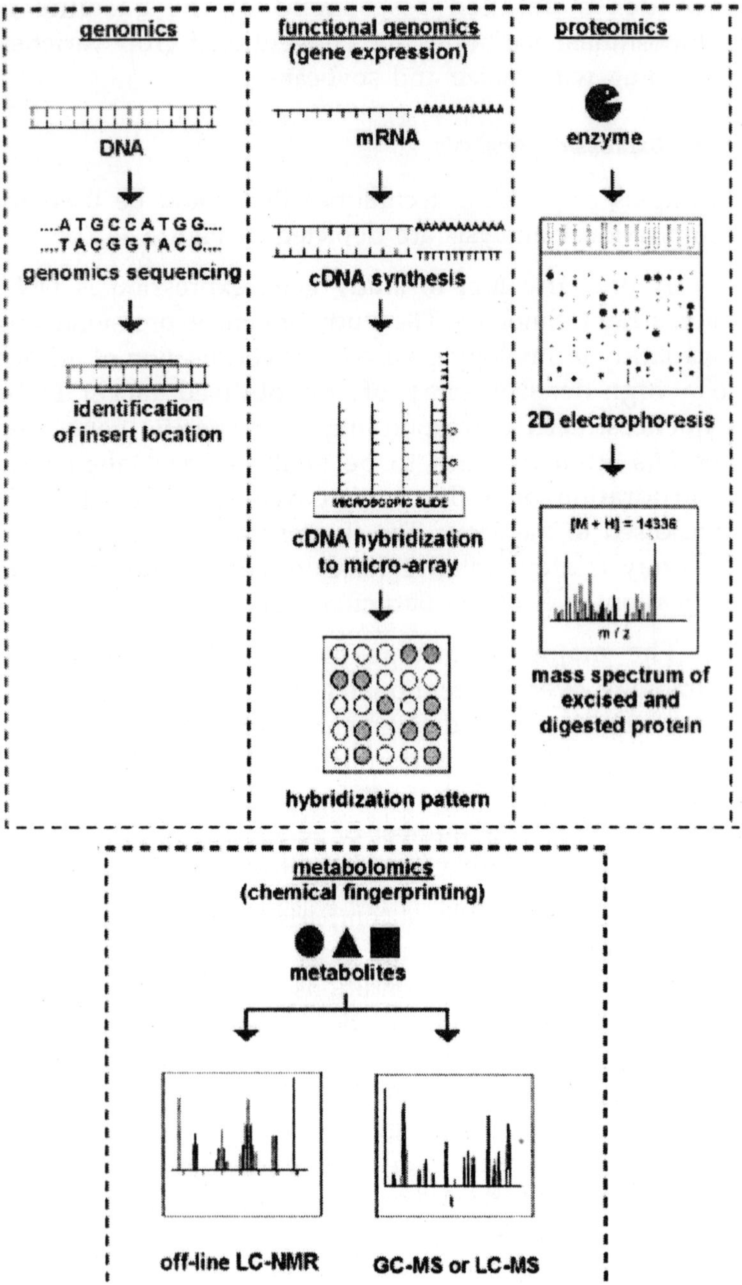
genomics
DNA
....ATGCCATGG....
....TACGGTACC....
genomics sequencing
identification of insert location
functional genomics
(gene expression)
mRNA
cDNA synthesis
MICROSCOPIC SLIDE
cDNA hybridization to micro-array
hybridization pattern
proteomics
enzyme
2D electrophoresis
[M + H] = 14336
m / z
mass spectrum of excised and digested protein
metabolomics
(chemical fingerprinting)
metabolites
off-line LC-NMR
GC-MS or LC-MS

Databases need to be established to generate information regarding the extent of natural variability with each of the new data points. In addition, profiling generates large data sets and appropriate software/hardware and statistical methods are needed to handle these. The potential value of transcript profiling for the safety assessment of GM food plants is currently under investigation using the tomato as a model crop.

To study differences in gene expression, 2 informative tomato expressed sequence tag (EST) libraries were obtained, one consisting of EST that are specific for the red stage of ripening and the other for the green unripe stage. Both EST libraries were spotted on the array, as were a number of functionally identified cDNA, selected on the basis of their published sequence. The array was subsequently hybridised with mRNA isolated from a number of different GM varieties, as well as with the parent line and control lines.

Preliminary results showed that different stages of ripening could be identified based on reproducible differences in gene expression patterns. Prospects are good that this method may be used effectively to screen for altered gene expression and, at the same time, may provide information on the nature of detected alterations (that is, whether these may affect the safety or nutritional value of the food crop under investigation). Additional studies are being carried out with GM potatoes and tomatoes in the framework of the EU sponsored project GMOCARE.

To be useful, the variability for each transcript needs to be established and knowledge gained on the relevance of each new assay point regarding safety and nutrition. Although these techniques may prove useful to identify differences among tissues or between a food component from a GM product and its conventional counterpart, the relevance to safety assessment will be challenging and has yet to be established. Therefore, these methods are not yet suitable for use in safety or nutritional assessments.

Applicability of the microarray technology to food safety assessment. A microarray customised for food safety assessment should preferably contain primarily cDNA or EST derived from metabolic pathways leading to the relevant nutrients and antinutrients, especially the natural toxins. Metabolic pathways for nutrients, antinutrients, and toxins (for example, glycoalkaloid biosynthesis) are poorly mapped, and, therefore, it is likely that, for the time being, the most important contribution of the microarray technology will be to help fill this gap in our knowledge of the physiology of crop plants.

Once the metabolic network is more fully elucidated, it is feasible that arrays could be constructed that can provide important information about changes in metabolic pathways that may need further investigation with respect to their implications for the food safety of the crop plant. Analysis of individual constituents can then, perhaps, be reduced to a limited number of proteins or metabolites that, based on the gene expression profile, seem to have been affected by the genetic modification. An acceptable ratio of the GM versus the parent line for individuals or groups of metabolites will then have to be established based on health implications.

The usefulness of this technology for the identification of unintended effects in GM crops depends largely on documented information about natural variations in gene expression levels in crop plants, which is still lacking.

Proteomics

Proteomics is an important research tool to study protein expression patterns. High-resolution, 2-dimensional gel electrophoresis can show those proteins present in a tissue which track within a given molecular weight and isoelectric point focusing range. New developments in mass spectrometry (MS) have increased the applicability of the technique.

Correlation between mRNA expression and protein concentrations is generally poor because rates of degradation of individual mRNA and proteins differ. Care must be taken in the extrapolation of these mRNA data to the proteomic and metabolomic level. Proteomics can be divided into 3 main areas:

— identification of proteins and their posttranslational modifications,

— "differential display proteomics" for quantification of variations in content, and

— studies of protein-protein interactions.

The method most of-ten used to analyse differences in protein patterns is 2-dimensional gel electrophoresis, followed by excision of protein spots from the gel, digestion into fragments by specific proteases, and, subsequently, analysis by MS (that is, peptide mass fingerprinting). It allows the identification of proteins by comparing the mass of the peptide fragments with data predicted by genetic or protein sequence information.

For crops with limited data on complete protein sequences, comparison with EST sequence databases may provide an alternative route toward elucidation of the putative function of an identified protein. The basic methods used to identify proteins separated by gel electrophoresis are electrospray ionisation and matrix-assisted laser-desorption ionisation time-of-flight (MALDI-TOF) MS analysis.

Electrospray tandem MS allows for fragmentation of selected ion species and subsequent analysis against a database. Some newer applications eliminate the need for separation of proteins from a complex mixture by gel electrophoresis. For example, a promising technique is affinity purification by isotope-coded affinity tags for more quantitative analysis, combined with multidimensional liquid chromatography and tandem MS.

Amino acid sequence data obtained from tandem MS data may also be useful for protein identification whenever no matches of single MS data with those of previously characterised proteins are found. When searching for unintended changes by 2-dimensional polyacrylamide gel electrophoresis (PAGE), the first step is to compare proteomes obtained from extracts of leaves or seeds of the GM plant and the closest genetic counterpart. If differences in protein profiles are detected, natural variations should be further examined by analysing the variation in levels of a number of different crop varieties grown under a number of different environmental conditions.

Detection of differences specifically related to the genetic modification process by analysis of general proteomes is difficult because of the many proteins that are not involved in such changes and the many changes that may occur because of different environmental conditions. If differences fall outside natural variations, identification of the protein is carried out, and this may lead to further safety assessment studies.

Proteomics can aid in understanding changes in cellular metabolism and physiology of food plants, regardless of the technology that was used to make the change. A major limitation in the use of proteomics for the detection of unintended effects is that changes due to the genetic modifications may not be easily distinguishable from changes due to environmental factors. Therefore, restrictive conditions for protein isolation, in addition to selection of proteins involved in important metabolic pathways, may reduce the number of confounding factors and yield more informative proteomes.

Further development of proteomics, combined with detection of specific proteins with antibodies or protein microarrays, may offer more effective ways to identify unexpected changes. The specificity of antibodies, for example, may allow for detection of specific proteins whose expression might have been influenced by a given genetic modification in GM plants.

In addition, validated and standardised protein extraction procedures have not yet been established. This is important because small variations in the many steps of sampling and extraction procedures may have a major influence on the resulting protein pattern. Characterisation and definition of ripening stages and storage conditions and of other genetic, agronomic, and environmental factors is essential because they all have a profound influence on the proteome.

Proteomics offers interesting possibilities for elucidation of cellular metabolic processes and their dynamics as influenced by breeding and selection processes and environmental influences. Identification of unintended effects and of toxicologically relevant proteins is one of its potential uses. Application of the technology for identification and assessment of unintended effects in GM crops is seriously hampered by the lack of information on natural variation in proteomes (for example, of commercially available lines) in addition to the abovementioned reproducibility and technical limitations.

Therefore, although interesting, the use of this approach will require considerable research and development before it is used for safety assessment. Recently initiated research projects funded by the UK Food Standards Agency specifically address this issue.

Metabolomics

Changes in mRNA levels or in proteins do not provide direct information about changes in biological function at the levels of metabolites in the food or feed components. A change at one level in a complex network does not necessarily lead to a particular change in function or phenotype. Therefore, these methods have limited direct value for risk assessment. Open-ended broad metabolite analysis may enable ways to define the biochemical function of plant metabolism.

A multicompositional analysis of biologically active compounds in plants (that is, nutrients, antinutritional factors, toxicants, and other compounds [the so-called metabolome]) may indicate whether intended and/or unintended effects have taken place as a result of genetic modification. The 3 most important techniques that are currently deployed for this purpose are gas chromatography (GC), high-performance liquid chromatography HPLC), and nuclear magnetic resonance (NMR). These methods are capable of detecting, resolving, and quantifying (in a relative sense) a wide range of compounds in a single sample.

For instance, profiles of isoprenoids using an HPLC method with photodiode array detection (PDA) were recently described for a GM tomato and Arabidopsis thaliana (Fraser and others 2000). Approximately 42 isoprenoids could be separated on a reversed-phase C30 HPLC system. The potential of GC to serve as a metabolomics tool for plants.

Metabolomics is being developed as a tool for comparative display of gene function. It has the potential to provide insight into complex regulatory processes and to determine phenotype directly. Researchers developed a GC-MS method allowing the quantification of 326 distinct compounds from Arabidopsis thaliana leaf extracts, with assignment of a chemical structure to approximately half of the compounds.

Four genotypes: 2 homozygous ecotypes and 2 single-point mutants from each genotype. Cluster analysis indicated that each genotype had a distinct metabolic profile, with the 2 ecotypes being more different than the singlepoint mutant compared to its parental ecotypes. Differences in profiles of soil or in vitro -grown tubers were observed. The results showed a remarkably low experimental variability (6%) compared to the much larger biological variability (20%).

Differences in the content of amino acids, citric acid cycle intermediates, and compounds indicative of osmotic stress not previously seen were found in GM crop material using this open-ended approach. In GM potato lines with increased yeast invertase or inhibited starch metabolism, differences were also observed compared to conventionally bred control lines. In the line with the invertase expressed in the apoplast, no elevated respiratory flux was observed, whereas in the line with the invertase expressed in the cytosol, appearance of phosphogluconic acid was observed, indicative of increases in components of the oxidative pentose phosphate pathway.

This open-ended approach of metabolomics provides the opportunity to identify unexpected changes, and, thus, insights into metabolic networks. Besides the metabolomic analysis of many classes of metabolites simultaneously, selective extraction, separation, and detection methods enable metabolomics of specific metabolite classes. In this way, more focused searches are possible into potential changes of (concentration/formation of) metabolites that may be caused by a particular genetic modification.

One example is metabolomics of isoprenoid compounds, including carotenoids, tocopherols, quinones, and chlorophylls, from chloroform extracts of plant tissues. For this purpose, highperformance liquid chromatography (HPLC) was used in combination with PDA detection, and UV-spectra of the separated compounds were generated. By applying this method to tomatoes in which an additional carotenoid biosynthesis gene had been inserted, the authors could determine which carotenoids had been altered or where a new formation had occurred.

Rice grains are characterised by a complex composition and large differences in concentrations of compounds. A method to fractionate the total rice extracts was developed enabling a GC analysis of a broad spectrum of major and minor constituents. The approach is based on consecutive extraction of lipids and polar compounds. Selective hydrolysis of silylated derivatives results in separate fractions of major (sugars) and minor (organic acids/amino acids) polar constituents.

Profiles of silylated/methylated compounds are obtained by means of gas chromatography - flame ionisation detection (GC-FID), and identification can be achieved by GC-MS. Further work will be carried out on GM rice varieties. It has been shown that the use of metabolomics techniques, such as off-line LC-NMR, may provide information on possible changes in plant matrices caused by variations in environmental conditions.

Metabolic profiles consisting of 1H-NMR spectra were obtained from different water and organic solvent extracts from GM tomato varieties, such as the antisense RNA exogalactanase fruit, and from their unmodified counterpart(s). Differences in concentration of low-molecular-weight components (MW <10 kDa) could be traced, by subtraction of the 1H-NMR spectra. To apply profiling methods on a routine basis, it is necessary to

— standardise sample collection, preparation, and extraction procedures;

— standardise and validate quantification;

— generate information on profiles of plant extracts and in particular on natural variations that are relevant to food and feed production (for example, crop lines cultivated in practice);

— further develop bioinformatic systems to treat large data sets;

— understand the quantitative limitations of each methodology;

— develop extensive databases to define the levels of natural variability;

— relate changes in gene expression levels to possible changes in protein and metabolite levels;

— evaluate the changes for their relevance to food safety; and

— select those changes that are relevant for food safety assessment.

Data analysis is of great importance, and univariate and multivariate statistics such as chemometric or pattern recognition techniques are necessary for the identification of relevant changes in metabolite patterns.

16

Technological Trends and Needs

Food safety is the assurance that food will not cause harm to the consumer when it is prepared and/or eaten according to its intended use. This definition is, however, hampered by the caveat that an absolute level of food safety cannot be obtained. It has therefore been recognised that instead, an acceptable level of risk has to be defined. Throughout the world, food production, preparation and distribution have become increasingly complex, and raw materials are often sourced globally.

Changes in food processing techniques, food distribution and the emergence of new food pathogens have changed the epidemiology of food-borne diseases. Food-borne microorganisms are continuously changing due to their inherent ability to evolve and their amasing capacity to adopt to different forms of stress.

New primary production technologies and food manufacturing practices are introduced all the time, food consumption patterns and the demographic structure of many countries continue to change. The implementation of Food Safety should be seen as an ongoing process, which is influenced by environmental, socio-economical, political and cultural factors.

Management Tools for Food Safety

End-product testing and microbiological criteria were traditionally used to evaluate the safety of a food product and to set criteria for deciding about rejection/acceptance of a food batch. This approach, however, has some limitations, especially for testing of pathogens that occur in low numbers and are non-randomly distributed. To guarantee the safety of foodstuffs producers have therefore shifted their focus towards the use of food safety management tools, most important HACCP (Hazard Analysis Critical Control Point), and the consequent application of hygienic measures, based on Good Manufacturing/Hygienic practice (GMP/GHP).

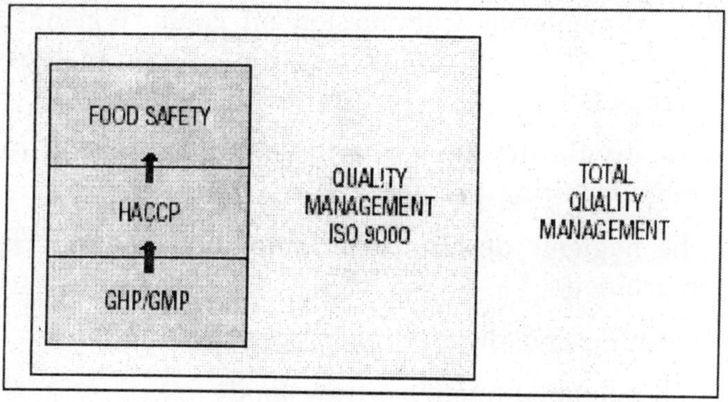

Figure 1. Food safety management tools

Food safety management tools use input of scientific information to identify critical contamination points in the food chain and the production process, and design measures to control them. However, the lack of reliable data is often limiting the usefulness of this approach and therefore data collection is one of the priorities for future food safety strategies. In the absence of relevant data, other strategies might still have to be used to control food hazards. The integration of different food safety tools is required for the production of safe foods from "field to fork".

As shown in Figure 1 quality assurance and safety tools can be combined in an approach that integrates quality control and quality assurance with a commitment to food safety.

Good Manufacturing Practice and Good Hygienic Practice

Good Manufacturing Practice (GMP) and Good Hygiene Practice (GHP) describe the basic measures that have to be applied during production, processing, handling and distribution, storage, sale, preparation and use. They form the basis for other approaches, most importantly HACCP. Requirements for GMP/GHP have been developed by the Codex Alimentarius Commission on Food Hygiene and the food industry, and cover the following general requirements :

— the hygienic design and construction of food manufacturing premises

— the hygienic design, construction and use of proper machinery

— cleaning and disinfection procedures

— the microbial quality of raw foods

— the hygienic operation of each process step

— the hygiene of personnel and their training.

Hazard Analysis Critical Control Point

The adoption of Hazard Analysis Critical Control Point (HACCP), based on GMP and GHP as the main approach to ensure adequate food safety has been accepted world wide. HACCP is a systematic, scientific approach to identify specific hazards and establish appropriate control systems. The focus is on the prevention of problems. The HACCP system does therefore not rely on end-product testing for pathogens. Instead, analyses are performed at points along the production chain, particularly to verify that Critical Control Point (CCP) are under control. Random sampling and analysis might be useful in the HACCP verification step.

Seven HACCP principles HACCP principles can also be applied to other parts of the food chain, under the condition that a critical control point can be identified.

Seven Steps of the HACCP -Plan

1) Hazard Analysis—Identify critical food-production steps where hazards might occur, assess their severity and human health risks, and determine preventative measures
2) Critical Control Points—Identify critical control points in the process at which the potential hazard can be controlled or eliminated Determination of
3) Specification of Critical Limits—Establish control measures and set up criteria to measure control at those critical points.
4) Monitoring—Monitor critical control points by establishing procedures for how the critical measures will be monitored
5) Corrective Actions—Take corrective action when the criteria are not being met
6) Verification—Routinely check the system for accuracy to verify that it is functioning properly and consistently
7) Documentation—Establish effective record-keeping procedures that document and provide a historical record of the facility's food safety performance

The introduction of the HACCP system allows regulatory agencies to shift their control efforts towards inspection of the HACCP system. The HACCP system can be combined with food safety objectives to develop quantitative process performance criteria.

Quality Management Systems

The aim of quality assurance is to ensure that a product continuously conforms to standard. The ISO 9000 series of Standards contains 5 different standards for quality management and quality assurance that can be used for total quality management and HACCP. They represent a generic management system which means that the same standard can be applied to any organisation irrespective of its products and sector of activity. The three core standards are: the ISO 9000 Quality management system - Fundamentals and vocabulary; the ISO 9001 Quality management system - Requirements; and the ISO 9004 Quality management system - Guidelines for performance improvement. The ISO 9000 were published in 1987 and completely renewed in 2000. In the future, ISO 9001:2000 will be the only standard through which an organisation can be certified.

Total Quality Management

Total Quality Management (TQM) is a structured system for satisfying internal and external customers and suppliers by integrating the business environment and continuous improvement. It aims at long-term success through customer satisfaction.

Risk Analysis

Risk analysis (RA) is a management tool for governmental bodies to define an appropriate level of protection and establish guidelines to ensure the supply of safe food. Risk in this context is defined as a health effect caused by a hazard in a food and the likelihood of its occurrence. Risk Analysis is useful to decide which hazards should be prevented, eliminated or reduced to acceptable levels. Risk analysis can provide the input to develop food safety policies that are based on scientific data to define acceptable levels of risks.

Food Safety Objectives can translate public health goals into quantitative targets to define appropriate levels of control. They are also important to evaluate the equivalence of different processes in respect to their safety performance. RA consists of three elements:

— Risk assessment is the use of scientific data to identify, characterise and measure hazards, assess exposure and characterise the risks involved with a food. In food safety regulations and policy development, risk assessment should consider the likely impact of a particular food safety problem and protective measures, and the urgency and controversy surrounding an issue. The final risk estimate should always contain information about assumptions made, and the degree of the variability and uncertainty in all steps of the risk assessment procedure. The Codex Alimentarius Commission has developed guidelines for risk assessment.

— Risk management is defined within the Codex as the process of weighing policy alternatives in the light of the results of risk assessment and selecting and implementing appropriate control options. The outcome of the risk management process is the development of standards, guidelines and other recommendations for food safety. If necessary, controls and regulatory measures are adopted.

— Risk communication is defined as an interactive process of exchange of information and opinion on risk among scientists, policy makers, and the public during the risk assessment and management process.

Food Safety Objectives

A Food Safety Objective (FSO) is a statement of a maximum frequency and/or concentration of a microbiological hazard in a food at the time of consumption that provides an appropriate level of consumer protection. For example, the health goal of reducing the incidence of a food borne illness by 50% from 10% to 5% per year could be translated into an FSO of 100cfu/g (colony forming units/ mass) as maximum bacterial number at the time of food consumption.

By projecting the possible growth of bacteria between production and consumption, a processing safety objective at the time of the production can be established. This can be used to define performance criteria for the production process, which would be the level of reduction that the process has to obtain to meet the processing safety objective.

Now the process criteria or product criteria that will be needed to achieve this performance can be defined. The process criteria might be time and temperature of a processing step, a product criterion might be the pH of a product. The critical limits of the process will be monitored using HACCP. The transition from microbiological criteria to Food Safety Objectives will also increase the choice for different methods of analysis.

Microbiological criteria have historically been used to accept/ reject food lots with unknown history at the port of entry and are subject to the same limitations as end-product testing. Microbiological criteria usually define o the microorgansim of concern and/or the toxins and the reason for concern related to these o the microbiological limits for the foodstuff at a specified point in the food chain o a sampling plan and a decision criteria to accept/reject, usually in the number of samples that have to conform to the limits o the analytical methods used for detection and/or quantification.

In 1985, the Food and Nutrition Board Council (FNB/NCR) addressed the subject of microbiological criteria and found that such criteria were of limited use. This is particularly true when safety assurance is the goal and therefore HACCP should be applied wherever possible. However, microbiological criteria are often demanded by governmental agencies, or between companies in a supply chain.

Microbiological Food Testing Limitations

Pathogens of concern often have a low incidence rate in a food-stuff or raw materials and are frequently not equally distributed. This results in statistical limitations for the usefulness of food testing to confirm or reject the presence of a pathogen. Not all food from a lot can be tested for the presence or absence of a certain pathogen. Instead, samples have to be taken, that should be representative for the whole lot. So called "lot-acceptance sampling plans" are often used to establish criteria for acceptance or rejection of a food lot and whether a product meets a certain set of specifications.

The number of samples and the % occurrence of the contaminant (% defective lots) determines the probability that a defective lot is accepted. For example, if 30 samples are collected from a lot with a defect rate of 1%, the probability for accepting is as high as 74%. The higher the prevalence of the contamination, the higher the probability to detect a defective sample. This is of practical relevance for many food pathogens.

E.coliO157:H7 in ground beef, for example, has a prevalence rate of less than 1%. Also the prevalence rate for listeria monocytogenes in ready-to-eat foods was reported to be between 1.08% to 4.91%. Non-random distribution of a pathogen adds to the problems of detecting a contamination when occurring at low rates. The effectiveness of testing is further influenced by sample preparation and the sensitivity and reliability of the analytical method.

The predictive value of a test is influenced by the prevalence of the defect, and additionally by the sensitivity and specificity of the test. With a test sensitivity and specificity of 95%, a pathogen with a high prevalence of 60% (e.g. campylobacter in poultry), the test will be positive in 96.6% of all cases were the sample is truly contaminated. This means that testing results will correspond well to the true level of contamination. With a test sensitivity and specificity of 95%, and a defect rate of 1%(e.g. E.coliO157:H7 in ground beef) the sample will be truly contaminated in only 16.1% of all positive test results.

The high rate of false-positive and false-negative results limit the value of testing in food with low defective rates. In this case, it can not be concluded that food that has been sampled for a certain pathogen and has tested negative is actually safe for consumption. This will also create an unnecessary high expenditure for testing with inadequate benefits in terms of increased food safety. As the defect rate of the product becomes low, it is more efficient to implement control systems that have been validated for the pathogens of concern.

These limitations should also be taken into account when methods for the detection of pathogens with low occurrence are developed. Food safety under these circumstances will be more efficiently accomplished using other tools than e.g. online testing.

Methods for Microbiological Testing

Safety and quality of a product are best controlled through effective management of those processing steps were hazards may arise. Microbiological contaminants in food are still the most important problems in terms of economical costs and epidemiological impact. Establishing critical limits within the processing steps and monitoring of the relevant parameters allows for control of the process. It is not possible to use microbiological parameters as critical control points, since the current methods are too slow to give the fast feed-back that is required within a HACCP system.

Even testing of raw materials and end-product testing is hampered by the fact that the methods are too slow. Therefore, much effort is put into developing methods which will allow the assessment of the microbiological quality and safety of foods in shorter time. Food is a difficult test matrix since it is extremely varied in its chemical composition, contains many different ingredients, and often contains an intrinsic microbial flora with varying amounts of shelf-limiting bacteria and even pathogens.

To account for the differences, the Association of Official Analytical Chemists (AOAC) recommends as many as 14 different food groups to be included in the validation procedure for Salmonella tests. Furthermore, food is processed with varying technologies and stored under different conditions. During processing bacteria might be damaged sublethally with subsequent problems of detection. As a result, different enrichment protocols might be necessary.

Microbiological methods are divided into traditional or conventional methods and rapid methods. Conventional methods are still considered the gold standard and are often required by national and international regulatory agencies as official control methods. The term Rapid Methods is used for different types of tests including miniaturised biochemical kits, antibody-based tests and nucleic acid hybridisation-based tests. They can be manual, semi-automated or fully automated. Ideally, the rapid methods should enable a quick estimation of the microbial parameters of the food to allow for immediate corrective actions during the manufacturing process.

However, the majority of rapid methods do not presently meet this demand. They do offer advantages in analysis time but also give the possibility to eliminate labour intensive steps and give the potential for automation. Due to the speed-up of the sampling process, sample treatment, detection/ enumeration procedure, the output of the laboratory can be improved. However, a real shortening in the time-to-result can only be obtained if the lengthy incubation procedures can be replaced by faster techniques or even omitted.

Traditional Methods

Traditional or conventional methods for isolation and identification of microorganisms involve the homogenisation of the food, its inoculation into selective media and incubation for a predetermined period. Samples can be solid, liquid, air or surface samples. The sampling methods depend on the material, the surface structure, and the expected contamination level.

Food samples are currently almost exclusively taken manually by destructive techniques such as excision or scraping. Diluting or enrichment steps might be necessary depending on the level of the microorganism. To substitute for the enrichment step techniques such as filtration, centrifugation, or magnetic separation might be employed. PCR can also be used for target amplification. After enrichment the broth is streaked on to solid media to yield isolated colonies that can be identified on the basis of their biochemical and serological characteristics.

Identification relies on cultural enrichment to increase the number of the target organism and to allow for resuscitation of injured cells. Followed by selective plating, these methods provide discrimination of the target organism from the background microflora but are non-enumerative. Enumeration methods are based on the ability of a bacterial cell to multiply in a nutrient-rich medium. Sometimes selective agents are added to favour the growth of a specific group of organisms, but most of these methods are quite unspecific. There have been a number of developments to speed up or automate the procedure, including the addition of colorimetric/fluorimetric substrates, the development of rehydratable nutrients eliminating the requirement to pour nutrient plates, and the development of biochemical identification kits.

Rapid Methods

Recently, a comprehensive review on "Rapid methods and Automation in Microbiology" has been written by Fung. An introduction to rapid methods can also be found in the Bacteriological Analytical Manual Online (BAM). Rapid methods can be used for isolation, early detection, characterisation and enumeration of microorganisms and their metabolites. According to Fung the development of methods can be described as follows:

- 1965-1975: the age of miniaturisation and diagnostic kits development
- 1975-1985: the age of immunological kit development
- 1985-1995: the age of genetic probes, molecular testing systems, and polymerase chain reaction (PCR) applications

— 1995-: the age of biosensor, computer chip and microarray systems.

Most rapid methods try to replace the selective and differential plating step of conventional methods with more rapid techniques. Pathogen testing requires very sensitive methods because for many pathogens in food there is a zero tolerance, namely that less than 1 cfu (colony forming unit) is allowed in two parallel 25g food samples. To obtain such sensitivity, all current methods need at least one enrichment step, usually incubation for a minimum of 6 up to 48 hours.

A new method must have high sensitivity, high specificity, high precision (repeatability), be rapid, robust and cheap. There is currently no method that will fulfil all requirements. Antibody-based methods as represented by the ELISA assay and DNA based methods as the polymerase chain reaction are the most widely used technologies in food diagnostics today. Also Immunomagnetic separation (IMS) is exploited in a number of commercially available kits and will become an even more important technology in the future.

While some microarray based systems are commercially available, most of the biosensors and microarray based applications are still on the level of prototypes. For other promising methods with future potential, but with less momentum for commercial development specifically in food diagnostics, as flow cytometry, bacteriophage technology or adenylate kinase, only a basic description of the method is given together with brief background information on the current stage of its development.

Viable cell count methods

One of the most common tests is the viable cell count, the enumeration of viable microorganisms. It gives information on food quality, food spoilage and food safety. Even if they cannot be considered to be a "rapid method", they will be covered due to their importance for routine food testing.

Originally, these methods were introduced to perform total viable counts, coliform counts, faecal coliform counts, yeast and mould counts, with the subsequent introduction of differential counts and pathogen counts. Although standard plate counts are easy to perform and cheap, they are time-consuming in terms of operation and data collection. Traditional total viable counts take three days to complete and utilise high amounts of reagents and consumables. For that reason, there is a considerable interest in the development of alternative methods for viable cell counts.

More methods based on growth and metabolic activity are available, like optical methods using specific chemical and physical changes connected with microbial growth that results in a colour change of the medium, turbidimetry, thermal methods and electrical methods based on impedance or conductance changes.

Microscopic methods for bacterial counts

A number of rapid methods using microscopic detection have been developed to employ bacterial counts that often can be performed in less than 1 hour. It is possible to investigate the total level of viable microorganism, specific bacterial groups or specific types of microorganism. These methods have already been utilised for counting bacteria in milk, milk products, water, beverages, raw meat, fish, poultry and food contact surfaces.

The Direct Epifluorescent Filter method (DEFT) is based on membrane filtration and detection either by fluorescent staining with acridine orange, which stains all cells, or by BAClight, which differentiates between viable and dead cells. The sensitivity of the method is 10 cfu/ml sample. The cells are than counted by eye or in semi-automated imaging systems. A similar method test can be completed within 20 minutes is currently implemented in Irish Food factories. Specific sample preparation methods have to be developed for each type of food. By using pathogen specific antibodies, specific microrganisms can be detected and the sensitivity of the method is about 10 cfu/ml sample.

The Chemunex Scan RDI system uses filtration and subsequent staining with a fluorescent dye, and analysis in a scanning chamber. The Microstar System uses a technology that traps bacteria in the matrix of the filter, with subsequent growth of colonies. By this method, only viable cells are counted. The filter is sprayed with a permeabilising reagent to release ATP, which can subsequently be detected using a bioluminescent agent and analysed using a CCD camera system.

Biochemical identification techniques

The identification of microorganisms with conventional methods using different biochemical test is labour intensive, with high usage of consumables and reagents. Miniaturised systems in the form of diagnostic kits have been developed that are based on the use of dehydrated growth substrates, or on ready-to use media. These techniques are usually termed modern biochemical identification techniques, are widely used and have a high importance for daily laboratory practice.

Pure cultures are grown in a variety of liquid or solid media and detected by a colour change, gas formation or enzyme-induced changes of the colour of the substrate. The results can be compared to a diagnostic chart, or analysed with a database. A number of diagnostic kits are on the market for the identification of pathogens, spoilage organisms, starter cultures, etc.. The systems offer visual analysis of the results or databases might be used to interpret and analyse the results in semi-automated and automated systems.

In a comparison with conventional methods the use of kits was considered accurate, efficient, time-, space-, labour-saving and cheaper than conventional methods. They have a widespread use and conventional methods are often mainly relevant for reference testing.

Antibody-based methods

Immunochemical analysis is a well-established technique with application to many different fields. Antibody-based methods have proven to be simple, rapid and sensitive for detecting and quantifying different types of food contaminants. Antibodies can be used to detect pathogens and spoilage organisms such as viruses, bacteria and moulds, but also low molecular weight food contaminants, such as mycotoxins, pesticides and veterinary drugs.

However, the development has been constrained by the availability of high performance antibodies against the selected target analyte. The specificity of an immunoassay is largely determined by the intrinsic specificity of the antibody used for detection. There are three different types of antibodies available for assay development: polyclonal antibodies, monoclonal antibodies and recombinant antibodies.

Recombinant antibodies are a very powerful new approach which offers many The selection of the appropriate antibodies allows the construction of tests with broad respectively narrow selectivity. Usually, an enrichment step is necessary. Tests can be single-use format like dip-sticks or done in miniplates, microplates or special formats. Only a few basic forms of antibody assay formats exist, but many different modifications are on the market.

Both homogeneous and heterogeneous assays are widely used, but homogeneous assays are considered superior in their ease-of-use. The different assay format include:

Latex-agglutination test: antibody-coated coloured latex-particles or colloidal gold particles are used for quick serological identification or typing of pure culture isolates of bacteria. With presence of the antigen, agglutination that can be visually inspected occurs. Reverse latex agglutination tests for soluble antigens and is often used for toxins. Immunodiffusion test format: an enrichment sample is placed in a gel-matrix with the antibody.

If antigen is present, a visible line of precipitation is formed ELISA (enzyme-linked immunoabsorbent assay) (Fig. 2). The ELISA test exists in different formats, but usually consists of a sandwich procedure.

The antibody is bound to solid matrix, usually a 96 well plate, but also dipsticks, paddles, pipette tips or other solid matrixes have been used. The sample is added, and upon presence of the antigen binding occurs that can be monitored using a secondary antibody coupled to an enzyme. Upon adding of substrate, a colour reaction occurs.

The principle of an ELISA is shown in Fig. 2. The detection limit of an ELISA is between 10^4 -10^6 cfu/ml, an overnight incubation to detect pathogens with low occurrence is therefore necessary. Many companies market systems that are based on ELISA assay formats. Recently, a number of fully automated systems have been introduced.

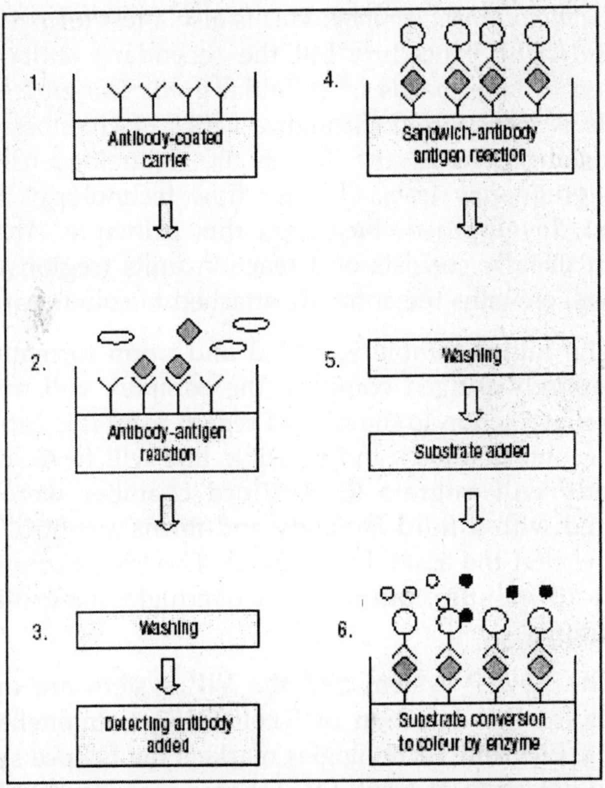

Figure 2. The test well is coated with the primary antibody against the targetanalyte

In figure 2 (1). The sample is added and upon presence, the analyte binds to the antigen and forms an antibody-antigen complex. Only the specific analyte is bound (2). The test well is washed several times to remove unbound antibody (3). The secondary antibody (detecting antibody) is added that recognises the antigenantibody complex. It is coupled to an enzyme for the detection procedure (4). Unbound antibody is washed away (5). The substrate is added. The Enzyme coupled to the secondary antibody converts the substrate into a coloured end product that can be visually inspected (6).

Immunochromatography: This is also a test format using a "sandwich" procedure but the secondary antibody is coupled to latex beads or colloidal gold. The enrichment sample is transported through a series of chambers, and no washing is necessary. This is the technology used for home pregnancy tests. "Lateral flow technology" is the newest development based on this principle. The test system usually consists of 3 reaction units (regions). The first well contains the antibody attached to colour particles.

The liquid sample is added and upon formation of the antibody-antigen complex, the complex will migrate by capillary action to the second region were it is captured by a second antibody and a visible line will form. Excess antibody will migrate to the third chamber were it is captured with a third antibody and forms a control band to show that the assay has worked. The test is completed within 10 minutes, but requires overnight incubation of the sample.

The Reveal system and the VIP system are on the market for the detection of E.coli O157, Salmonella and Listeria. Eichrom Technologies markets the Eclipse system for the detection of E.coli O157. Some recently developed immunoassays make use of the growth of the pathogens in the actual assay. An example is the Unique Salmonella detection system developed by Tecra. An overnight incubation is also necessary for this method.

The sample is added to the first tube and a dipstick coated with Salmonella antibodies is added for 20 minutes to capture Salmonella and transferred into a second tube were it remains for incubation to allow further growth of the captured bacteria. Newly produced bacteria will also be captured by the antibodies. The dipstick is then placed in 3 more tubes for reaction with the secondary antibody, washing and the colour reaction. The system can be used in laboratories with small volume throughput.

Immunomagnetic separation

Immunomagnetic separation (IMS) techniques can be used to replace or supplement and speed up the enrichment step that is usually necessary before the detection of pathogens. IMS differs from the assay format described above, as it is not a detection method in itself. It can be combined with different end-detection methods.

For IMS, paramagnetic beads coated with various antibodies are used to concentrate target bacteria selectively. For example, two companies currently produce antibody-coated magnetic particles for the use in food diagnostics: Dynal Biotech and Vicam. Dynal beads are uniform, super paramagnetic polystyrene microspheres. The hydrophobic surface allows the absorption or coupling of different molecules.

Paramagnetic and superparamagnetic beads only have magnetic properties when exposed to a magnetic field and can therefore be easily resuspended without aggregation in the absence of a magnetic field. For the separation procedure the sample is mixed with the antibody-coated paramagnetic particles, and after incubation, the bead-bacteria complex is extracted through application of a magnetic field with a magnetic device. The complex is washed once or twice.

Bacteria remain viable after IMS and do not need to be detached from the beads prior to cultivation. The procedure is not really quantitative due to differential recovery rates of the bacteria-bead complexes and the possibility of several bacteria attaching to one bead. Recovery rates range typically between 25 to 50%. An advantage is the reduction of the background microbial flora that shortens the time required for selection of suspect colonies for confirmation.

Under optimised conditions pre-enrichment times as low as 4 to 6 hours allow the detection of 1-2 bacterial cells/g of the original food sample. There are a number of commercial available test kits that combine IMS with various detection methods. The EHEC-Tek system uses IMS to capture E.coliO157:H7 prior to the ELISA detection procedure, which leads to enhanced assay sensitivity.

The Origen assay system is based on immunomagnetic electrochemilumenescence. For the PATHIGEN test available for Salmonella, Listeria, Campylobacter and E.coli O157, the technique is used in a sandwich immunoassay format. A pathogen specific antibody is immobilised on the magnetic particle. The second antibody is labelled with a light-emitting compound, the ORI-TAG label. With the pathogen present, both antibodies bind to it and form a complex which is then transported to a flow cell. A magnet captures the beads on the surface of an electrode in a flow cell.

The label ($Ru(bpy)_{32}+$) and a coreactant (T.PA) are oxidised at the surface of the electrode. The chemical reactions produce an excited label that emits light which is measured by a photodetector. The amount of the pathogen correlates to the intensity of the emitted light and determines the concentration of the pathogen present in the sample. The measurement takes less than 1 minute. A pre-enrichment of about 6 h is still necessary. This technique is thus a real one-shift method with test results available in less than 8 hours. The sensitivity of the system is about 1-2 cfu/g of the initial food sample.

IMS used in combination with PCR has several advantages. Often, PCR assays which are directly applied to food have a significantly decreased sensitivity which is based on the unspecific inhibition of the DNA polymerase by different food components. Through removal of these inhibiting food components from the sample, PCR ready DNA is easier obtained and the sensitivity of the PCR assay can be improved.

Moreover, PCR assays use very small sample sizes of 1 to 20 µl. IMS allows the concentration of the sample by a factor 5 to 10 and thereby increases the sensitivity of the assay. A technique called magnetic capture-hybridisation PCR (MCH-PCR) involves lysing the bacteria to release DNA and hybridisation with pathogen- specific gene sequences using biotin labelled DNA probes. Following capture of the hybrids by streptavidin coated magnetic particles, the bound DNA is used for PCR amplification.

A disadvantage of the IMS based techniques is the need to optimise for each type of sample since the performance of the method is determined by a number of factors e.g. the concentration of coated antibody per mg particle, the number of particles per test, the incubation conditions, type of sample and washing regimes, and the antibody-specificity. Under optimum condition, a recovery rate of 90% has been reached. IMS has proven to be a useful tool to enhance the speed and sensitivity of different detection methods. All of the steps have currently to be performed by hand, but an automation of the method can be expected in the next future.

Assays based on Nucleic acid

Nucleic acid-based assays can be either performed directly, or after amplification of the target sequence. Polynucleotides with a sequence complementary to the single-stranded DNA or RNA are used as probes. DNA sequences that are homologous to all members of the family of interest, or unique sequences that are particular to a species can be chosen for the construction of the probe.

Probes that target different genera, species or different strains can thereby be constructed. Genes associated with virulence are often used for detection of pathogens. The ribosomal RNA represents an attractive target for probe design since it is present in multiple copies in most organisms, and databases with sequences for several thousand microorgansims are available. Ribosomal RNA contains stretches of conserved sequences interspersed with variable sequence regions, providing the scope to design single-stranded DNA probes for broad-range or specific target detection.

The binding of the probe to the complementary cellular DNA/RNA is called hybridisation. In order to detect hybridisation, the probe needs to be labelled. Among others, Biotin, Digoxigenin and different fluorochromes have been used for labelling. Since there are many different assay formats available which are produced by many different manufacturers.

For detection of food pathogens, the GENE-TRAK system uses direct hybridisation in a sandwich hybridisation format. The capture probe directed to specific rRNA sequences is linked to a solid support and is designed to hybridise with the target sequence. A second probe is used that hybridises to a sequence adjacent to that site. This probe is labelled with fluorescein on both ends. The hybridisation product is detected with secondary antibodies labelled with horseradish peroxidase that can be visualised after a chemical color reaction. A number of other manufacturers produce similar tests.

Amplification-based methods

Molecular assays have become faster, less expensive and through the development of homogenous assays easier to use, making them more attractive for routine use in food laboratories. There are a number of methods available that allow the amplification of the target sequence or the detection signal, including branched DNA technology (bDNA), Nucleic Acid Sequence-based Amplification (NASBA), Qâ Replicase, the Ligase Chain Reaction (LCR),

Transcription- mediated amplification, strand displacement amplification (SDA) and rolling circle amplification (RCA).

In a recent Tekes technology review a short introduction into methodology and an overview over the diagnostic companies developing these methods can be found. While the PCR method has been most widely accepted, other assay formats that have yet not reached practical applications in food diagnostics, have been developed. For a recent review see also One promising method is the Rolling Circle Amplification (RCA) method that uses both signal and target amplification. The technique allows direct amplification on a solid phase, and is therefore suitable for the combined use with microarrays.

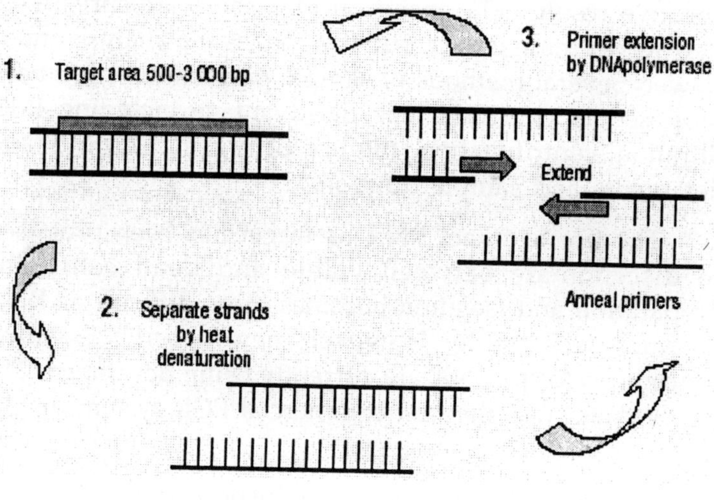

Figure 3. The Polymerase Chain Reaction.

The Nucleic Acid Sequence-Based Amplification (NASBA) amplifies RNA and DNA targets as single-stranded antisense RNA. Reverse Transciptase, Rnase H and T7 polymerase and two primers are needed. PCR is now an accepted method to detect pathogens by amplification of the target DNA and detecting the target PCR products (Fig. 3).

Detection of the amplification product has previously been done by gel electrophoresis, which is not a very convenient technique for routine use in a food laboratory. Recent advances have led to the development of homogenous assays for real-time detection which allow greater sensitivity and easy detection of PCR products that make PCR more suitable for routine-use in food laboratories.

A major disadvantage of the PCR technology is that it will detect both viable and non-viable bacteria. The decreased sensitivity of the PCR reaction directly applied to a food sample is well known and can also lead to false-negative results. The inclusion of a short pre-enrichment step solves these issues, but also makes the analysis much slower. Several methods of sample preparation have been proposed to overcome this disadvantage including IMS, filtration, centrifugation, the use of organic solvents and detergents and sample dilution.

The use of intact RNA as target has been proposed to discriminate between viable and dead cells. The preparation and isolation becomes more difficult as RNA is much less stable. However, new specific isolation procedures for m-RNA are currently being developed. The BAX-system was the first commercial PCR system for the detection of food-pathogens.

The FSIS (Food Safety and Inspection Sevice) USA has recently adopted this system for screening of ready-to-eat food, meat poultry and egg products for Listeria monocytogenes. The BAX-system works directly from an overnight enrichment of the target organisms without a separate DNA isolation step. All PCR reagents required for the test are incorporated into a single tablet. The principle of the TDF-PCR detection is based on an intercalating dye, SYBR green I, that binds to the double-stranded PCR product.

The assay is an in situ process, simultaneously amplifying the target DNA and directly detecting the increasing fluorescence signal during the annealing/elongation phase of the PCR reaction. One additional thermal cycle is run consisting of a denaturation step and a product annealing step. Fluorescence is monitored directly in the PCR reaction tubes during this final cycle. The rate of increase in fluorescence over time during the product annealing step yields a characteristic pattern for positive samples that can be differentiated from negative samples.

The TaqMan system also allows data evaluation directly in the reaction vessel, and through online detection of the amplification reaction a quantification of the amount of target sample present in the DNA probe. The probe is an oligonucleotide with both a reporter fluorescence dye and a quencher dye attached. When both dyes are attached to the probe, reporter dye emission is quenched. During each extension cycle, the Taq polymerase cleaves the reporter dye from the probe, which subsequently emits fluorescence. Online following of the production of the amplification product allows quantification of the amount of DNA in the original sample.

The Molecular Beacon Technology uses a hairpin-shaped hybridisation probe. The probe contains a reporter fluorescence dye and a quencher dye attached to respective ends of the probe sequence. In the absence of the target PCR products the beacon remains in a hairpin shape and the fluorescence is quenched. During the generation of target PCR products, the beacons will attach to the PCR products and cause the hairpin molecule to unfold, and fluorescence will be emitted, allowing realtime detection of target PCR products.

In contrast to the TaqMan system, the amplification product is measured directly. Using several different fluorochromes, several PCR products can be followed in the same reaction tube. Figure 4. Show the principle elements of a biosensor. Binding of an analyte to a recognition element leads to a signal change at the transducer. Numerous different components or technologies can be used at the various stages of detection.

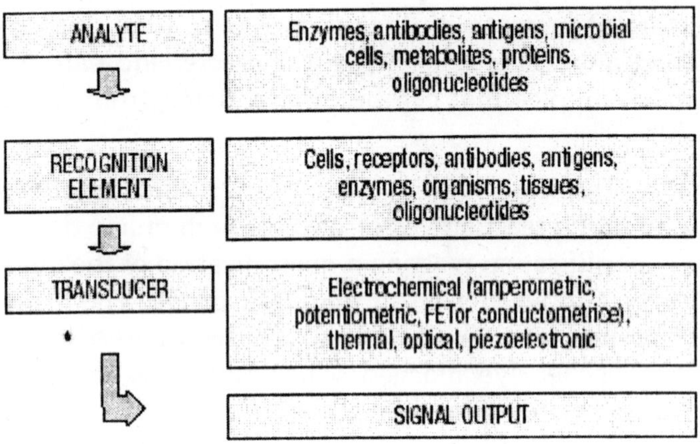

Figure 4. The principle elements of a biosensor.

Sensor Technology

The most promising breakthroughs of the development of on-line or on-site, sensitive, low-cost, rapid methods for routine-use are expected to be made in the area of sensor technology. Many prototypes for food diagnostic application in the food and drink industry are currently being developed. They have high potential for automation and allow the construction of simple and portable equipment for fast analysis. These properties will open up many applications within quality and process control, control of fermentation processes, quality and safety control of raw materials, and for HACCP monitoring.

Sensor Technology covers a wide area of diverse techniques, including opto-chemical sensors and biosensors. Biosensors are a subgroup of chemical sensors where the analytical devices are composed of a biological recognition element such as enzymes, antibodies, receptors, proteins, oligonucleotides, or even a whole cell coupled to a chemical or physical transducer.

A transducer measures the changes that occur when the sensor couples to its analyte. The sensitivity of the system is determined by the type of tranducers employed. Biosensors can be used for the detection of very different analytes such as pathogens, pesticides and toxins. Biosensors can be grouped according to their biological recognition element into immunosensors using antibodies and hybridsensors using DNA or RNA probes.

Microarrays are constructed of a high number of parallel hybrid receptors and can therefore be considered biosensors. For the construction of low density microarrays suitable for the use in food diagnostics, electrochemical transducers have proven to be a cheap and efficient choice. Electrochemical transducers include potentiometric, amperometric, piezoelectric, capacitive and conductive instrumentation.

Optical immunosensors are most popular for bioanalysis and optical transducers comprise the largest group of transducers used for biosensors today. Optical transducers include fibre optic transducers and those without fibre optics as spectrophotometers, fluorometers and luminometers. Due to the diverse nature of the technologies used in the field, we will focus on a description of microarrays and optical sensors and include some highlighting examples of commercial systems and prototypes available for applications within food diagnostics.

However, only a limited amount of methods are combined and currently exploited for their use in food diagnostics. As recognition elements, bioaffinity based receptors that use the selective interaction between ligand and receptor, antibody or nucleic acid are most widely used. As transducers, electrochemical and optical systems have gained practical importance.

Microarrays

Microarrays often are also referred to as biochip, DNA chip, DNA microarray or gene array, and allow to conduct many analyses in parallel. The size and complexity of such arrays varies considerably. High-density microarrays (up to 100.000) are mostly used for research, as gene expression profiling and microbial genotyping, and drug development purposes.

The use of microarrays as research tool was described in a recent review. Medium (some hundreds to thousands of different analytes) and low-density microarrays (some 10 to hundred different analytes) are more suitable for mass production and can therefore be applied for diagnostic purposes due to their considerably lower cost. The immobilised molecule is termed the probe, while the solution-phase molecule is termed the target. Probes can consist of cDNA, oligonucleotides, proteins and are immobilised in a regular arrangement on the surface of a solid substrate.

As solid substrates, a. a. glass, nitrocellulose, nylon, polypropylene, and silicon have been used. Probes can be deposited on the surface of the substrate by deposition of presynthesised probes using specially developed printing systems (microspotting) or by in situ synthesis at individual locations on the array. To do an analysis the unknown sample is incubated with the microarray, allowing the hybridisation of complemen-tary sequences.

For high density microarrays fluorescent probes are mostly used for the detection of the hybridisation reaction. For medium and low density microarrays, electrochemical detection offers a cheap and sensitive detection system.

An alternative assay format are the so called virtual arrays. Beads are used for the attachment of the probes. Since the beads have no fixed location, the identity of the probe attached to the bead has to be encoded differently. Luminex Corporation uses polystyrene as the matrix material, with 100 spectrally distinguishable beads analogous to a 100 spot microarray. GeneScan Europe offers a microarray for the detection of gene manipulated organism. The first version of the GMO Chip contains a total of 14 tests and analyses.

The GMO Chip Kit can recognise DNA from respectively: plants and virus species, commonly used construction elements in the field of genetic engineering, or specific genetically-engineered modifications (for identifying approved and non-approved GMOs). A prototype based on a low-density microarray with electrochemical detection is currently being developed for the assessment and quality control of yeast based bioprocesses. The device consists of a low-density microarray allowing parallel detection of 16 probes.

The microarray can be used for the detection of spoilage organism in the brewery process. The sensitivity of the method is 106-107 DNA copies. An additional application is possible through the investigation of the gene expression in yeast with microarrays. By investigating characteristic changes in the expression of certain yeast genes, critical stages in the fermentation process can be identified and used for better control of the fermentation process. Also the quality of the yeast for the brewery process can be analysed with this method.

Optical Biosensors

Optical biosensors have been developed for rapid detection of contaminants in foods, including pathogens, and several have evolved into commercial prototype systems. Optical transducers have gained from advances in optical fibre and laser technology. They are based on methods such as bio- or chemiluminescence, reflectance, scattering, and refractive index, and UV-vis absorption.

The basic principle utilises electronic or optical transduction technology to monitor a parameter of the reaction between a biomolecule and an analyte into a quantifiable electrical or optical signal. This signal can be related to concentration. In Fibre-optic biosensors a fibreoptic probe serves as the transducer. The analyte in the food interacts with the bioactive molecule, usually an antibody. Antibodies can be immobilised directly on the fibre, either on the blunt end or along the sides of a fibre tip. The binding of antibody and analyte is detected as a change in an optical signal measured through the fibre-optic assembly.

The light from a laser travels to the fibre tip and penetrates into the area outside the tip. A fluorescently labelled complex binds to the antibodies on the tip. The fluorescent signal then radiates in all directions, and some of it travels back up the fibre tip to the detector. Non-selective adsorptions and interferences due to background fluorescence have been the major drawbacks for the development of this sensor type. So far, only a few commercial systems are available on the market.

Additionally, some optical biosensor prototypes have been developed. The first commercial applications have been obtained with the surface plasmon resonance (SPR) sensor. Biacore International AB, technology offers this assay for determination of water-soluble vitamins and for veterinary drug residues, antibiotics and growth promoters. The SPR technology monitors biomolecular interactions in real time.

The detection principle relies on surface plasmon resonance, an electron charge density wave phenomenon that arises at the surface of a metallic film when light is reflected at the film under specific conditions. In Biacore systems, SPR is used to monitor molecular interactions occurring on a biospecific surface; the changes in molecular weight as molecules bind. An increase in molecular weight on the surface coating of the sensor chip causes a corresponding increase in refractive index which alters the angle of incidence required to create the SPR phenomenon (the SPR angle), which is detected as a response signal.

The system is based on a microfluidic device in which the analyte flows past the detector surface. Detection of molecules in solution can be made either by direct binding to the biosensor coating molecules, or by competition binding with soluble capture molecules added together with the sample. Future applications might include protein quality and the detection of allergens genetically modified (GM) proteins, BSE prions, pathogens and biocide residues Another SPR-based biosensor system has been developed by Texas Instruments and packaged into handheld sensors for small molecular analytes that permit on-line and real time analysis.

The antibodies are coupled to urease and ammonia is produced upon binding of the antigen in the presence of urea. The ammonia gas induces measurable changes of the optical properties of the sensor using a laser detection system. This optical biosensor has been tested for the detection of 12 different bacterial species and has a sensitivity of 500 cfu/ml sample. Upon binding of the fixed antibodies, the liposomes rapture, thereby releasing dye or other markers.

The Raptor is a portable, automated fibre-optic biosensor that is commercially available from Research International. Four different antibodies are immobilised on separate fibre probes. Binding to the appropriate probe generates a fluorescent signal within a few minutes. The assays are fully automated for simple operation and data analysis.

Flow Cytometry

Flow cytometry has mostly been used as a research tool for the study of microorgansims. However, the development of cheaper instrumentation should make it more useful for routine applications in the future. Flow cytometric studies of microorganism are based on their staining with a fluorochrome. Stained cells in suspension are injected into a fast-moving carrier fluid. The cells arrive in a single file in the measuring area where a light source is focussed on the fluid stream.

The individual cells pass the light beam very rapidly. The stained cells will be excited and emit light at a specific wavelengths, which will be measured by a detector. It is possible to analyse the cells using staining with different specific fluorochromes or to monitor their light scattering properties. A particle will deflect or absorb light in a characteristic way depending on its size and structure and the angle of the light. Multiparameter analysis with up to eight different parameters is possible with certain instruments.

Flow cytometers can also be equipped with cell sorting devices allowing to sort out cells with a certain profile. By this method, staphylococcus aureus could be separated form e.coli to a purity of 95%. The technique is called FACS (Fluorescence activated cell sorting). When staining cells for analysis, it is possible to use nucleic acid stains, protein stains, antibody and dyes for membrane potential measurement and enzymatic activity. The technique can provide rapid information on live and dead cells and information on their physiology and enables both qualitative and quantitative analysis of microbial cells in liquids. Due to the complexity and inherent cost of the equipment, the practical use of the method is still limited to research. However, with the appearance of cheaper systems on the market, it is considered to be a promising technology for the future.

Bacteriophage-based Techniques

Bacteriophages are viruses that selectively infect a certain range of bacterial hosts. This property which allows discrimination between different bacterial strains has been exploited in "phage typing" of bacterial samples. Recently, a new method for the detection of food-borne pathogens has been developed were the bacteriophages are engineered to carry the luxAB gene, expressing the enzyme luciferase. The infected cell will produce light that can be measured with a photodetector.

Viability is essential for the light production, so that the assay can discriminate between viable and non-viable cells. The assay is more rapid due to the faster replication cycles of the virus compared to bacteria, generating a 20 to 100 fold increase of the virus particle within 30 minutes. At present, there are no commercial kits available using this technology.

Hygiene and Environmental Testing in Food Industry

The hygiene of all surfaces, machinery, pipes and instruments have an impact on the quality and the safety of the food produced. The microflora of the processing plant is composed of microorganisms from the air, water, raw materials, dust, dirt, and people. Food can be contaminated during processing due to cross contamination via contaminated raw materials, personnel, aerosols, malfunctioning or improperly cleaned equipment, and other major contamination sources in the plant.

Hygienic design and proper cleaning have therefore a direct effect on the quality of the food produced. Work to provide design standards is being performed by the European Hygienic Equipment Design Group (EHEDG), the 3-A Committee, International NSF, the European Committee for Standardisation (CEN), Safety in Biotechnology Standards, and the International Standards Organisation (ISO). The Codex Alimentarius Committee on Food Hygiene has produced several hygiene standards.

Hygiene testing or environmental monitoring involves testing for possible contamination of working surfaces to monitor the success of the cleaning procedures. One big problem in the food industry is the formation of biofilms on a number of different surfaces. Biofilms can build up on any surface in the food industry. Amongst the most contaminated surfaces are floors, drains, and some conveyor belts. Less contaminated surfaces are walls, ceilings, and the equipment. For a surface to be colonised by bacteria, conditioning is needed. This occurs when a wet surface forms and soluble molecules concentrate on the surface.

In the food industry spontaneous conditioning occurs through adhesion of food molecules and/or cleaning agents and disinfectants. Biofilms develop when microbes attach to these conditioned surfaces and start to secrete a protective extracellular matrix containing polysaccharides and glycoproteins. The adhesion process seems to trigger a number of characteristic genotypic and phenotypic changes, so that bacteria in a biofilm differ significantly from bacterial cells in solution. These changes might also underlie their high resistance to all antimicrobial agents.

In addition, biofilms will occur under unstable environmental conditions exposed to different types of stress as fluctuating access to nutrients, chemical shock and desiccation. The coprotection towards other forms of stress that often is acquired in the process might additionally protect from those chemicals used in hygiene operation. Several commercial alkaline cleaning products were shown to detach 10-90% of a biofilm grown under laboratory conditions but did not produce a significant decrease in the number of adhering cells in a naturally grown biofilm.

Removal of biofilms through cleaning is very difficult, but cleaning will effectively reduce fresh contaminations from pathogenic microorganisms through external sources and prevent establishment of those microorganism in biofilms. The hygienic design of the plant and equipment is hence the basic tool to prevent the establishment of biofilms. Dryness of surfaces and air help to efficiently inhibit the establishment of biofilms.

Biofilm pieces that occasionally detach might be sources of contamination for a food product with pathogens. This contamination will be difficult to detect due to the sporadic and random occurrence of the process. Fast regrowth of the bacteria in the biofilm due to only partial removal after the cleaning cycle leads furthermore to a decreasing quality of the food product. For that reason, fast and robust methods are needed to monitor cleaning efficiency and to establish efficient sanitation operation procedures. As for microbial methods within HACCP system, those methods need to be fast, and should ideally provide results in real time.

Hygiene Monitoring and Rapid Methods: Hygiene testing is currently based on conventional cultivation after sampling of the surfaces using contact plates or swabbing. Comparison between microscopic evaluation of biofilms and culturing shows an order of magnitude difference in the total number of the organisms. Due to the strong adherence of the biofilms it is necessary to use substances that induce detachment. SprayCult from Orion Diagnostica is a product that is used for the detachment of biofilms.

ATP Bioluminescence: The method is based on the firefly luciferase reaction which emits light in the presence of ATP. This phenomena is called bioluminescence. ATP occurs in all living cells and is therefore useful to indicate the presence of microbial organisms and food residues. The presence of ATP on a surface can therefore be used to indirectly assess the cleanliness of a surface. There are several ATP monitoring systems on the market. They involve swabbing of the tested surface using a surfactant and measuring the presence of ATP after addition of Luciferin and Luciferase in a luminometer. The reaction is very fast and provides readings within 2 minutes.

The major limitation of the method is the relatively low sensitivity. According to theoretical limits, a reliable ATP measurement should be obtained with at least 1000 bacterial or 10 yeast cells in the sample. A further limitation is the swabbing procedure which is used for sampling, and which might only remove 1 to 10% of the contaminants from a surface. To address that problem the BioProbe system consist of a luminometer connected to a suction cup which allows measuring directly at the surface. Currently there are quite many companies offering ATP instruments.

Adenylate kinase: The limited sensitivity of the ATP bioluminescence method might be overcome in the near future by an assay developed by DERA Porton Down that is using adenylate kinase rather than ATP as a cell marker. Adenylate kinase is a constitutively expressed enzyme that catalyses the reaction

$$Mg^{2}+ \ {}^{*}ATP + AMP \leftrightarrow ADP \ {}^{*}Mg^{2}+ + ADP$$

This reaction serves to rephosphorylate the AMP produced in many reactions. ADP can be further phosphorylated to ATP. However, by adding excess ADP, the reaction can be driven in the other direction to produce ATP, that can be measured using the luciferin/luciferase reaction. The reaction creates about 40 times more ATP for the bioluminescence reaction than would be possible for the ATP naturally present on its own.

The Adenylate kinase (AK) assay offers therefore a 10 to 100 times higher sensitivity than the ATP bioluminescence assay. At its present state of development at DERA Porton Down the AK assay has been shown to have a detection limit of 10 E.coli cells within 5 minutes. AK technology is simple to use and does not require the development of new detection principles or instrument systems. Since the AK content of bacterial cells is much more constant than their ATP content the new method yields results that are directly comparable between different species. The AK assay can be made both highly specific and ultra-sensitive when used in combination with specific binding techniques such as solid phase capture by antibodies. Kits are to be expected in the near future.

Riboprinting and pulse-field gel electrophoresis: Riboprinting and Pulse-field gel electrophoresis are used for the genetical characterisation of microbial cells. These methods are able to produce characteristic DNA finger printing patterns that can be used for surveillance programmes for food borne diseases. By comparing patterns collected in central databases can be connected to sources of contaminations can be traced even for geographically separated incidents.

With the same technique it is also possible to trace sources of contamination in a food production plant.Afully-automated Ribo Printer Microbial Characterisation system is available that can identify organism to genus, species and subspecies level automatically (DuPont Qualicon, Wilmington, DE, USA). The use of these methods is currently limited to research laboratories. Many contract research organisations (CROs) offer the application of these methods for detection of contamination sources as a service to the food industry.

The field of rapid methods is growing fast with numerous analytical options being available for different testing situation. These new rapid methods have the potential to offer considerable advantages regarding their sensitivity, rapidity and potential for automation, allowing increased sample through-put for analytical laboratories, and ease-ofuse. Most of the current rapid methods detect single target, but more and more methods with multiflexing properties are developed. Rapid methods are well suited for screening purposes for large numbers of food samples. Since most methods lack the required sensitivity, culture enrichment is still necessary before analysis. Benefits of enrichment include dilution of inhibitors, recovery of sub-lethally damaged cells and differentiation between viable and non- viable cells.

Validation Process

The aim of a validation procedure is to ensure that a new method performs according to specified requirements. It always contains an evaluation of the specificity, sensitivity, precision and accuracy and kit performance data of the new method. The validation process is usually coordinated by an independent validation body, which also has to develop the rules for this process.

The actual validation is performed by different laboratories, in the form of peer-reviewed and collaborative laboratory studies. The validation of a rapid method has to demonstrate that it produces results that are comparable to the results obtained by a reference method. There are different validation and standardisation bodies as ISO (International Organisation for Standardisation), CEN (European Committee for Standardisation, AOAC (Association of Official Analytical Chemists,) NMKL (Nordic Committee on Food Analysis), and AFNOR (Association française de normalisation).

AOAC is the validation body that is responsible for the validation of rapid methods in the U.S. NordVal and MicroVal are the nordic and European bodies for the validation of rapid methods for microbiological analysis. A validation procedure offers the advantage of that the diagnostic company can prove the performance of their new method in comparison to well established methods that are known by the user. This is very valuable information to the user for the decision process of which method to use for his particular purpose.

Conversely, a lack of validation data when introducing a new method necessitates that more tests have to be performed by the food laboratory, requiring time and systematic work effort. It is obvious that this adds to the costs of establishing a new technique in a food laboratory, and moreover, not all laboratories will have the means to perform the necessary studies at all.

A good validation procedure implies that a method has acquired a certain degree of standardisation before it is implemented into practical use in food laboratories. It is not an easy task to maintain the reliability of a standard while following the ever-increasing pace of technological and scientific development. This is clearly an inherent problem to this area. For official control there is a need to maintain and rely on certain established standards to be able to guarantee the quality of the control, but this will invariably hamper technological progress and innovation.

On the other hand, technology developers need to create markets for their new products quickly to be able to create incentives for further developments. The validation process also offers important information to official food control agencies, and allows a degree of control over the quality of the method being offered on the market. The performance results from the validation process can be used by regulatory agencies to accept a new technique as an official testing method. This will have a positive influence on the penetration of rapid methods, since they currently cannot be used as official control methods. Clearly this does not increase the incentive to adopt new technologies.

With the existence of numerous different validation bodies, and in the absence of a mutual recognition, most diagnostic companies that strive for validation of a method will have to deal with several separate regulatory agencies. In most cases no new laboratory studies will be required, and the various agencies might just end up cross validating each other. This will not lead to any added value for the method in question, but will increase costs and prolong timelines.

Moreover, it will decrease confidence of the end-user in the whole validation process, if no common acceptance standards can be created. A validation process with mutual recognition between different validation bodies resulting in global acceptance would be a major step towards the opening of a common market for food diagnostics. In practice, the major problem with validation procedures is that they require a very significant effort from the laboratories involved, measured both in time and absolute workload.

Currently, the AOAC approves 12 to 18 official methods annually. This indicates that the capacity for regulatory acceptance procedures of validation is limited and can easily become a bottleneck. New tools are needed to guide the selection and evaluation of new rapid methods, and to ease exchange of experience with various new test methods.

MicroVal is a European certification organisation for the validation and approval of alternative methods for microbiological analysis of food and beverages. The MicroVal procedure aims also at the recognition of validated rapid methods as official control methods. The lack of a common validation procedure was perceived a major obstacle for the establishment of rapid methods by both the diagnostic companies and the food industry.

An Eureka project was started to develop a certification scheme describing the methodology and the organisation to be used. The interest form both diagnostic companies and manufactures for a European validation for rapid methods is also reflected in the composition of the general committee (MGC):

— *Manufacturers*: Organon Teknika (UK), Laboratories 3M Santé (France), R. Biopharm (Germany), Diffchamb (Sweden)

— *Users*: Nestle (Switzerland), Unilever (the Netherlands), UNIR (France)

— *MicroVal third parties*: AFNOR (France), DIN Certco (Germany), Campden & Chlorleywood Food res. Ass. (UK).

MicroVal has recently started its activities in form of a pilot project. The target is to conclude a validation procedure within one year.

NordVal is the Nordic committee for the validation of alternative microbiological methods, including analyses of food, water, feed, animal faeces and food environmental samples. They have so far evaluated 8 different rapid test kits. The validation procedure is aimed at the acceptance as official control method, but is not implicitly built into the validation process. A secretariat will be established at the Institute of Food Safety and Toxicology within the Danish Veterinary and Food Administration.

The *AOAC International* (Association of Official Analytical Chemists) is an independent, non-profit organisation that focuses on the coordination of the development and validation of analytical methods in the area of analytical chemistry and microbiology. The AOAC official methods have been incorporated into the FDA methods. Accepted methods are published in the AOAC periodicals Journal of AOAC and Inside Laboratory Management and The Referee.

Currently, the AOAC approves 12 to 18 official methods annually. Compared to the amount of methods and test kits developed it is becoming increasingly obvious that this is not enough to keep pace with the fast development of rapid methods. To fill that gap, a new project was introduced in June 2002 called "e-CAM" (electronic-Compilation of Analytical Methods). "e-CAM" consists of a collection of information about methods in a database available on-line.

The idea is to collect all the available information on a method in one database to ease the selection of the best suitable methods for a user. It also aims to promote exchange of experience with certain methods and make that information more readily available. Different validation status are considered such as regulatory methods, harmonised collaboratively (fully) validated methods, multiple-laboratory validated methods, single laboratory validated methods, and research non-validated methods with some information on their quality.

Food Diagnostics Market

The market size for food diagnostics can be assessed through different estimates about the number and types of microbiological tests, but no direct statistical data are available. As to be expected, the estimates differ depending on how the market was segmented, and some of the data are not completely congruent. The market data show that traditional methods still clearly dominate the market of routine test for quality control, with different forms of total plate counts as the most performed test. Rapid methods are still a comparably small market.

The market penetration of rapid tests is greater in the U.S. than in Europe. The world market for food analysis totals •1.1 billion. This includes testing for pathogens, spoilage organisms, chemical contaminants and process chemical tests (mostly different chemical analyses of food compounds). The world market for all microbiological tests is around •1.1 billion, with 755 million microbiological tests performed in 1998. 56% of those tests were performed on food, 10% on beverages, 30% on pharmaceuticals, and 4% for environmental testing.

Food microbiological tests created global sales of around •565 million, and an estimated 440 million food tests were performed in 1999. In the US sales for food microbiological testing were about •209 million compared to •170 million in Europe in 1999. North America and Western Europe together make up about 2/3 of the world market. Table 1 shows the global sales of traditional food microbiology tests versus rapid methods by region. With an annual growth of 9% the world market for food microbiology is expected to reach •864 million in 2005.

Table 1. Worldwide sales of traditional food microbiology tests versus rapid methods by region.

1999	Traditional Methods			Rapid Methods		
Region	Sales (• millions)	Market Share (%)	Growth per year (%)	Sales (• millions)	Market Share (%)	Growth peryear (%)

North America	140	35	5	63	55	16
Western Europe	115	29	4	23	20	27
Latin America	61	15	8	15	13	17
Japan	56	14	7	12	10	20
Rest of the world	28	7	4	2	2	12
Total	400	100		115	100	

Whereas the market share for rapid methods is 31 % (compared to total sales) in the US, in Europe only 17 % of the sales come from rapid methods. This shows the smaller market penetration of rapid methods in Europe compared to the US. Media and reagents still make up the largest share with •350 million or 62% of the market. In comparison, rapid methods generate global sales of •115 million or a 20% total market share. ATP methods for hygiene monitoring have a market share of 9% and generate •50 million of sales (Table 2).

Table 2. Worldwide food microbiology sales by technology.

Year	1999			2005	
Technology	Sales (• millions)	Market Share (%)	Growth per year (%)	Sales Share (%)	Market (• millions)
Media/ID reagents	350	62	5	460	53
ID systems	50	9	5	65	8
Rapid Tests	115	20	19	244	28
ATP Hygiene	50	9	15	95	11
Total	565	100	9	864	100

Within rapid methods, immunoassays still stand for 60 % of all the tests sold and generate sales for •70 million, followed by DNA methods (PCR) methods with •25 million and a 22% market share (Table 3).

Table 3. Worldwide food microbiology sales by rapid methods.

Technology	Sales(• millions)	MarketShare (%)
ELISA/EIA	70	61
DNA	25	22
Impedance	10	9
Flow Cytometry	5	4
ATP	5	4
Total	115	100

Globally, 370 million enumeration tests were performed which correspond to 84% of all food microbiological tests and a market value between •185 and 280 million. Forty four million tests for pathogens were performed worldwide in 1999. Most of the pathogen tests (35 million) were done using traditional methods, while only 9 million tests were performed using automated ID methods or rapid methods. 90 % of all pathogen tests are made for 5 food borne pathogens: Salmonella, Listeria, E.coli, Campylobacter and Staphylococcus. The world market for these tests accounts for •220 million, with rapid tests generating a sales volume of •45 million.

The European market for food diagnostics totals 490 million • with rapid methods amounting to •86 million (Table 4). In Europe around 240 million tests worth •220 million for spoilage microorganisms and 28 million pathogen tests to a value of •98 million test are performed annually.

Table 4. European food analysis/diagnostics sales.

	(• million)	(• million)	%
		Rapid methods	Rapidmethods
Food pathogens	98	13	15 %

Spoilage microorganisms	220	38	17 %
Chemical contaminants	53	20	40 %
Process chemicals	120	16	10 %
Total	491	87	

Rapid tests for pathogen detection generate sales volumes of around •13 million, corresponding to 13% of the market. For detection of spoilage microorganism rapid tests generate sales of around •38 million, which corresponds to 17% of the market. The market size for chemical contaminants as mycotoxin, food allergens, hormones, antibiotics is •53 million in Europe, with rapid tests representing 40% of the market or •20 million.

Genetically Modified Organisms

The world market for GMO (Genetically Modified Organisms) is treated separately from the food microbiology market. Even if similar methods are used and developed for GMO testing, such as PCR, ELISA and DNA microarrays, the testing market is influenced and shaped by different factors.

In Finland to date no products are on the market labelled "containing GMO" or "GMO free", respectively. The world market for GMO testing is still relatively small; the global market was •21 million in 1999, with sales worth only •3 million in Europe, corresponding to a 14% market share. 62% of the market for GMO tests consist of immunoassays, mostly rapid dipstick tests for qualitative yes/no results at grain elevators as grain is prepared for shipment in the U.S.

The costs for sampling are in the range of •2-10 for rapid tests based on immunoassays and between •100-300 for highly sensitive PCR tests per sample. Growth rates in the range of 100% were predicted for DNA based tests and the market is estimated to grow to •70-100 million in 2005. GMOtesting is supposed to guarantee that the raw material and the end-products are free of gene-modified protein or DNA. The current requirement through "The Novel Food regulation" in the European Union requires labelling of food products containing more than 1% products of gene-modified origin.

The market is currently still sorting out developing with regard to methods of verification and price premiums for non-biotech foods. This has led to the import of certified GMO-free food to Europe. Both domestic and international marketing functions in the U.S. have begun to differentiate markets for biotech and non-biotech products and this has created a new business opportunity. It is estimated that 24% of the grain elevators will separate biotech corn and 20 % will segregate biotech soybeans. The European market is willing to pay a price premium for non-biotech food. In a survey, 53% of the European consumers stated that they would be willing to pay more for non-GMO foods.

In July 2001, the EU commission adopted two proposals; new regulations on GMO traceability and the labelling of GMOs and products produced from GMOs, and a proposal for a regulation in food and feed. The new labelling system will additionally require labelling of:

— All foods produced from GMO irrespective of whether there isDNAor protein ofGMorigin in the final product.

— All genetically modified feed.

Currently no labelling requirements are in place for feed or for highly refined soy or corn oil that are free of traces of protein or DNA. If these proposals are implemented, problems will arise with labelling required for foods which contain neither DNA nor gene-modified protein and therefore cannot be tested for it at all to control compliance to regulations. The whole testing concept might be put into question if this caveat is not adequately addressed.

Also with feed products some problems are to be expected, since separation cannot be maintained without significantly higher costs. In reality, food containing GMO, even if only in trace amounts, has long been present on our tables. The estimate is that 70% of all processed foods contain ingredients derived from corn or soy and often they serve as basis for secondary ingredients that are not listed on the label. The future of labelling requirements and the market for GMO testing are therefore highly uncertain.

Bibliography

F. Pudil, H. Wijaya, V. Janda, J. Volfova, H. Valentova and J. Pokorny. "Changes in Citrus hystrix oil during autooxidation". *In Food Flavors; Formation, Analysis and Packaging Influences,* E.T. Contis et.al (eds), p. 707. Elsevier Science B.V., England, 1998.

Giyatmi, J. Basmal, C. H. Wijaya and S. Fardiaz. "Effect of fungi and fermentation period on the wuality of dried fish stick (Katsuobushi)", *Journal of food Technology and Industry,* 11(2): 10-20.

Kamaruddin Abdullah and H. Wijaya. "Production of Safe Food to Meet the Global Need and Market Indonesian Experience". J. Of International Society for Southeast Asian Agricultural Sciences) ISSN 0859-3132, Vol. 6 (June): 54-57, 2000.

Morr C. V., *Whey and Whey Products Available for Utilisation,* in Whey and Lactose Processing, edited by Zadow, J.G. Elsevier Applied Science, London, 1992.

Muchtadi, C. H. and C. H. Wijaya. "Food Technology Development in Indonesia with Emphasis on Food Additives Application and regulation". *Challlenges in Food Processing and Packaging,* May 1997, p. 8, 1997.

United Nations Environment Programme Industry and Environment (UNEP IE), *Food processing and the environment.* UNEP Industry and Environment 18(1). ISSN 0378 9993, 1995.

Wijaya, C.H., "Synthesis of garlic compound vinyll-dithiin and its derivatives, and their anti-funga determination using bioautography methode", *Journal of Food Technology and Industry,* 11 (1) : 48 -59, 2000.

______________, Apriyantono, A. "Aroma volatiles of several unique tropical fruits and spices. In : Flavour Research at The Dawn of The Twenty-First Century". *Proceeding the 10 Weurman Flavour Research Symposium,* Beanue , France 24-28 Juni 2002, ed. J-L, Le Quere, P.X. Etievant. Lavoisier and Intercept, Londres- Paris - New York pp. 749-756, 2003.

Wirakartakusumah, M.A. and C.H. Wijaya. "Recent Developmentof Food Flavors' in Indonesia". *Proceeding of the Third Malaysian Conference of Essential Oils and Flavor Chemicals.* Langkawi , Malaysia . July 20-23, 1992.

Index